W O R K B O O K
to Accompany
Woodworking

Second Edition

Nancy Macdonald

 CENGAGE

Australia • Brazil • Mexico • Singapore • United Kingdom • United States

Workbook to Accompany Woodworking,
Second Edition
Nancy Macdonald

Vice President, Careers and Computing:
Dave Garza

Director of Learning Solutions: Sandy Clark

Sr. Acquisitions Editor: James DeVoe

Managing Editor: Larry Main

Sr. Product Manager: Jennifer Starr

Editorial Assistant: Aviva Ariel

Vice President, Marketing Programs:
Jennifer Ann Baker

Director, Market Development: Deborah Yarnell

Sr. Market Development Manager: Erin Brennan

Sr. Brand Manager: Kristen McNary

Senior Production Director: Wendy A. Troeger

Production Manager: Mark Bernard

Content Project Manager: David Barnes

Senior Art Director: Bethany Casey

Cover Image: ©Shutterstock.com/Mike
McDonald; Wood grain images: © C Miller Design

For product information and technology assistance, contact us at
**Cengage Customer & Sales Support, 1-800-354-9706
or support.cengage.com.**

For permission to use material from this text or product, submit all requests online at **www.cengage.com/permissions.**

ISBN-13: 978-1-133-94962-6

ISBN-10: 1-133-94962-2

Cengage
20 Channel Street
Boston, MA 02210
USA

Cengage is a leading provider of customized learning solutions with employees residing in nearly 40 different countries and sales in more than 125 countries around the world. Find your local representative at: **www.cengage.com.**

Cengage products are represented in Canada by Nelson Education, Ltd.

To learn more about Cengage platforms and services, register or access your online learning solution, or purchase materials for your course, visit **www.cengage.com.**

Notice to the Reader

Publisher does not warrant or guarantee any of the products described herein or perform any independent analysis in connection with any of the product information contained herein. Publisher does not assume, and expressly disclaims, any obligation to obtain and include information other than that provided to it by the manufacturer. The reader is expressly warned to consider and adopt all safety precautions that might be indicated by the activities described herein and to avoid all potential hazards. By following the instructions contained herein, the reader willingly assumes all risks in connection with such instructions. The publisher makes no representations or warranties of any kind, including but not limited to, the warranties of fitness for particular purpose or merchantability, nor are any such representations implied with respect to the material set forth herein, and the publisher takes no responsibility with respect to such material. The publisher shall not be liable for any special, consequential, or exemplary damages resulting, in whole or part, from the readers' use of, or reliance upon, this material.

Printed at CLDPC, USA, 01-22

Contents

SECTION VII: FINISHING

SECTION VIII: SPECIAL TOPICS

PART II: SKILL SHEETS

Preface

This workbook is designed to accompany *Woodworking,* second edition, and is intended to provide a multitude of exercises to practice what you have learned in the book, as well as to evaluate your progress through the skills and techniques associated with constructing new projects.

It includes two sections:

Section I—Practice Questions

Each chapter within this section consists of various question types to check your learning as it relates to the key concepts presented in the corresponding chapter in the book. These questions include multiple choice, completion, identification, and short answer. Complete each set of questions following your study of the chapter material. Review any material that you miss. Your instructor may also choose to assign as classwork or homework, and require you to submit for grading.

Section II—Skill Sheets

Each of these sheets is based on a Procedure outlined in the book, and allow an instructor a method for evaluating your skills in the lab environment. Each skill sheet is based on a standard template, and includes information that you need to complete the skill, a scale for evaluating your accomplishment of each step in the skill, and page references for the book if you need to further study a particular procedure. These sheets are designed to be completed by the instructor and utilized for grading purposes.

Along with the author, we at Cengage Learning encourage you to take advantage of the exercises in this workbook to help lead you toward a pathway of success in this trade.

Enjoy your experience as a woodworker, and we wish you luck!

PART I
PRACTICE QUESTIONS

CHAPTER

1

The Woodworking Industry

MULTIPLE CHOICE

Identify the choice that best completes the statement or answers the question.

_____ 1. Planes, saws, and hammers were first used in _____ and look very much like the same tools we use today.
   a. ancient Rome
   b. England
   c. the Middle Ages
   d. the Industrial Revolution

_____ 2. Early settlers needed wood for building homes, barns, business establishments, and the ships that transported goods. They built mills powered by _____ to grind grain for flour and to turn logs into lumber.
   a. geothermal energy
   b. slave labor
   c. water wheels
   d. solar energy

_____ 3. The invention of the circular saw, which no carpenter today can imagine working without, is credited to _____.
   a. the ancient Romans
   b. the trade guilds in Western Europe
   c. someone watching the action of a windmill
   d. the Shakers

_____ 4. During the Middle Ages in Europe, a young man would be sent out to train with a master at the age of 7 or 8, and typically worked for him for a period of 7 years. He was called a(n) _____.
   a. novice
   b. apprentice
   c. cadet
   d. journeyman

3

_____ 5. During the Middle Ages in Europe, the _____ were above the journeymen.
- a. bosses
- b. superintendents
- c. foremen
- d. masters

_____ 6. Today, more than _____ people are employed in one capacity or another in working with wood.
- a. 10,000
- b. 100,000
- c. 500,000
- d. 2 million

_____ 7. The U.S. Department of Labor's _____ is a good resource you can utilize from your computer to learn more about careers in various categories.
- a. Bureau of Labor Statistics
- b. Bureau of Engraving
- c. Bureau of Mines
- d. Treasury Department

_____ 8. The U.S. Department of Labor's _____ gives you information on the nature of the work, working conditions, employment, training and advancement, job outlook, earnings, related occupations, and sources of additional information.
- a. Student Career Handbook
- b. Census
- c. Income Statistics Report
- d. Occupational Outlook Handbook

_____ 9. _____ is a national organization serving high school and college students who are enrolled in technical, skilled, and service programs.
- a. The YMCA
- b. The Masons
- c. VISTA
- d. Skills USA

_____ 10. The first thing a person notices about you is _____.
- a. the way you talk
- b. the way you look
- c. the kind of briefcase you carry
- d. the way you shake hands

COMPLETION

Complete each statement.

1. The first tools were made of _____ and bone, which were often fastened to wooden handles.

2. By the 1600s, the white pine of the Northeast was highly prized by _____ for masts.

4

3. The advent of steam power and later _____ made it easier to process lumber, and allowed lumber to be processed much faster.

4. It is possible today for an operator using "_____ numerical controls (CNC)" to program a machine to perform a sequence of operations automatically.

5. During the Middle Ages, an apprentice gained _____ status at the end of his apprenticeship and upon demonstrating his skills.

6. The word "craft" comes from the old English word *craft,* meaning "_____."

7. A(n) _____ is an organization of wage earners formed for the purpose of serving the members' interest with respect to wages and working conditions.

8. A(n) _____ is supervised, practical training undertaken by a student or recent graduate.

9. A(n) _____ is a brief account of one's professional or work experience and qualifications designed to give potential employers a snapshot of who you are and what you can offer to their company.

10. A(n) _____ is a formal meeting in person arranged for the assessment of the qualifications of an applicant.

IDENTIFICATION

Identify each item pictured below. Write the letter of the best answer on the line next to each number.

a.

© Cengage Learning 2014

b.

© Cengage Learning 2014

c.

© Cengage Learning 2014

d.

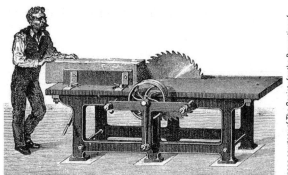

e.

_____ 1. plans being drawn using drafting software

_____ 2. table saw

_____ 3. installing hardwood floors

_____ 4. completed deck

_____ 5. pit saw

SHORT ANSWER

1. Explain what a water wheel is and how it works.

_____

_____

_____

_____

_____

_____

_____

2. Explain the process of a successful job interview.

_____

_____

_____

_____

_____

_____

_____

_____

3. Identify several traits of a good employee.

_____

_____

_____

_____

_____

_____

_____

4. Explain why the way that you leave your job is important.

_____

_____

_____

_____

_____

_____

_____

5. Identify some ways of "getting ahead" in the workplace.

_____

_____

_____

_____

_____

_____

_____

_____

_____

_____

_____

CHAPTER
2
General Safety Practices

MULTIPLE CHOICE
Identify the choice that best completes the statement or answers the question.

_____ 1. If you cannot keep your attention on what you are doing, you should _____.
   a. use dangerous tools only during daylight hours
   b. not operate dangerous tools
   c. use dangerous tools only at night
   d. have an assistant who will warn you of dangers

_____ 2. A simple auxiliary jig for holding a small piece while boring on the drill press is a hand-screw _____.
   a. clamp
   b. push stick
   c. shield
   d. chisel

_____ 3. If you wear corrective glasses in the shop, they should be _____.
   a. shatter-resistant
   b. lightweight
   c. plastic
   d. set in a wraparound frame

_____ 4. Wood dust has been linked to certain types of _____, such as lymphoma.
   a. asthma
   b. allergies
   c. cancer
   d. bacterial infections

_____ 5. If you work with _____, you must be particularly careful to clean up when you finish working for the day, because the combination of wood dust, moisture, and sufficient oxygen can lead to spontaneous combustion.
   a. respirators
   b. pine
   c. sawdust
   d. green wood

_____   6. Sweeping the shop is not enough. You must periodically _____ the accumulated dust.
a.   vacuum
b.   wash
c.   scrub
d.   recycle

_____   7. Do not overreach while in the shop because you must be _____ at all times.
a.   confident
b.   creative
c.   relaxed
d.   securely balanced

_____   8. You are more likely to get hurt with a _____ tool.
a.   sharp
b.   dull
c.   clamped
d.   clean

_____   9. When making adjustments to tools, you should _____.
a.   disconnect them from the power source
b.   keep the power on
c.   connect them to a generator
d.   lower the power level by half

_____ 10. You should not work when you are under the influence of _____ drugs.
a.   prescription
b.   legal
c.   illegal
d.   any

COMPLETION

Complete each statement.

1. The number one safety rule is to stay _____.

2. _____ are safety devices designed for use with specific tools, and they keep the operators' hands from getting dangerously close to spinning blades, bits, and cutters.

3. A(n) _____ is a purchased or shop-made device that makes a job safer and easier.

4. A(n) _____ is a shaped wood or plastic device that allows the operator to move material past the blade without putting his fingers into the danger zone.

5. When you open to a chapter on safety, protecting your _____, ears, and lungs is probably what comes to your mind first.

6. Good hearing protectors will screen out _____ frequencies, which cause the most damage, but they will still allow you to hear normal conversation.

10

7. A person who is regularly exposed to 110 decibels for a period of more than one _____ risks permanent hearing loss.

8. Wood scraps and dust can also present a fire hazard, so your shop should be equipped with an approved _____ in the event of fire.

9. A(n) _____ kit should contain, at the minimum, splinter tweezers, needles for splinter removal, povidone-iodine solution, latex gloves, instant ice packs, clean plastic bags, an asthma inhaler, an eye cup and eye wash, Band-Aids of various sizes, 4" × 4" gauze pads, sterile rolled gauze, adhesive tape, butterfly bandages, and sharp scissors.

10. To cut down on the time you spend picking splinters out of your hands, keep a pair of _____ in your tool kit, and do not hesitate to put them on when moving material in bulk.

IDENTIFICATION

Identify each item pictured below. Write the letter of the best answer on the line next to each number.

a.

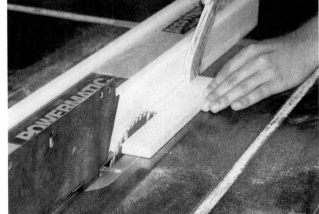

© Cengage Learning 2014

b.

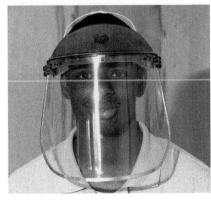

© Cengage Learning 2014

c.

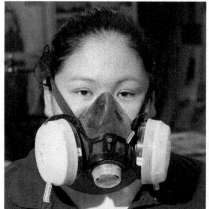

d.

e.

_____ 1. a simple auxiliary jig

_____ 2. face shield

_____ 3. push stick

_____ 4. ear muffs

_____ 5. respirator

12

SHORT ANSWER

1. What are the greatest hazards of woodworking?

_____

_____

_____

_____

_____

_____

_____

_____

2. When is using a tool guard not necessary?

_____

_____

_____

_____

_____

_____

_____

_____

3. What are the decibel levels of some common sounds?

_____

_____

_____

_____

_____

_____

_____

_____

4. How should you dress while in the shop?

_____

_____

_____

_____

_____

_____

_____

_____

5. What are the general guidelines for working safely in the shop?

_____

_____

_____

_____

_____

_____

_____

_____

14

CHAPTER

3

Hand Tools

MULTIPLE CHOICE

Identify the choice that best completes the statement or answers the question.

_____ 1. Good steel tape measures are available in lengths ranging from 12 to _____ feet.
  a. 15
  b. 20
  c. 25
  d. 30

_____ 2. A *square* is a T-shaped or L-shaped tool used for drawing and testing _____ angles.
  a. acute
  b. right
  c. obtuse
  d. straight

_____ 3. A marking _____ gives you a very accurate, fine line that will not be smudged.
  a. knife
  b. compass
  c. square
  d. plane

_____ 4. Tooth spacing is referred to as *TPI,* or _____.
  a. teeth placed inside
  b. teeth placement inset
  c. teeth per incision
  d. teeth per inch

_____ 5. The _____ is the gap created as the saw is used.
  a. dust
  b. kerf
  c. set
  d. TPI

_____ 6. A _____ may be defined as a metal tool with a sharp beveled edge that is used to cut and shape wood.
   a. plane
   b. backsaw
   c. chisel
   d. hammer

_____ 7. _____ planes are usually the first plane a beginning woodworker acquires.
   a. Block
   b. Bench
   c. Shoulder
   d. Scrub

_____ 8. _____ differ from files in that they have individually raised, triangular-shaped teeth.
   a. Drawknives
   b. Card scrapers
   c. Rasps
   d. Spokeshaves

_____ 9. The three parts of a screwdriver are the handle, the shaft, and the _____.
   a. screw
   b. bore
   c. tip
   d. hammer

_____ 10. A relatively new type of screw head is the _____ head.
   a. square
   b. Phillips
   c. flat
   d. slotted

COMPLETION

Complete each statement.

1. Steel tapes and wooden _____ are the woodworker's primary measuring tools.

2. Compasses made specifically for woodworking are called _____.

3. The _____ saw is designed for making cuts with the gram; the cutting edges of its teeth have a flat front edge and are not angled.

4. A(n) _____ is a tool used to flatten, smooth, and reduce the thickness of wood.

5. _____ planes differ from bench planes in that the cutting iron is embedded with the bevel up, and they do not have a chipbreaker.

16

6. The _____ drill is a small, portable drill that is operated by turning the handle; it is sometimes called an egg-beater drill.

7. A(n) _____, also called a *drawshave,* is a knife fitted with a handle at each end of the blade; it is used with a drawing motion to shave material from a surface.

8. Scrapers cut by means of a(n) _____, which is a sharp hook of metal that is turned on the edge of the scraper by burnishing it with a steel rod.

9. A(n) _____ is a type of hammer with a softer head than the metal head found on hammers.

10. There are many different types of screw heads, but the oldest and most familiar are slotted and _____.

IDENTIFICATION

Identify each item pictured below. Write the letter of the best answer on the line next to each number.

a.
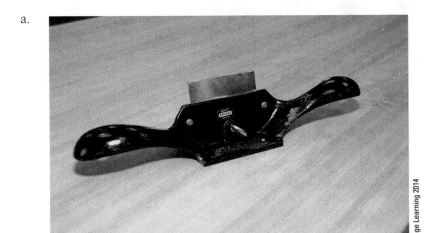
© Cengage Learning 2014

b.
© Cengage Learning 2014

c.

d.

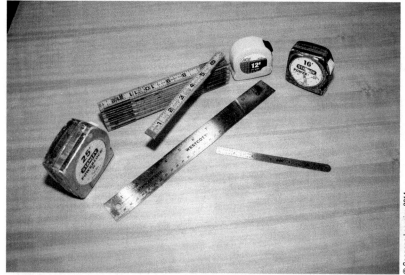

e.

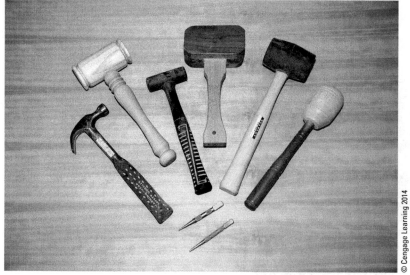

_____ 1. Tapes and rules

_____ 2. Bevel squares and protractor

_____ 3. Hammer, mallets, and nail sets

_____ 4. Cabinet scraper

_____ 5. Chisels and tip protectors

SHORT ANSWER

1. What are the parts of a combination square, and how is a combination square used?

_____

_____

_____

_____

_____

_____

_____

_____

_____

2. What are the steps you should follow when starting a cut with a handsaw?

_____

_____

_____

_____

_____

_____

_____

_____

_____

_____

3. How can you use a chisel to cut a hinge recess?

_____

_____

_____

_____

_____

_____

_____

_____

_____

4. What is the procedure for setting up and using a bench plane?

_____

_____

_____

_____

_____

_____

_____

_____

_____

_____

5. What is the difference between a hammer and a nail set?

_____

_____

_____

_____

_____

_____

20

CHAPTER

4

Portable Power Tools

MULTIPLE CHOICE

Identify the choice that best completes the statement or answers the question.

_____ 1. The _____ is probably the most widely recognized and used of the portable cutting tools.
   a. circular saw
   b. miter saw
   c. power planer
   d. router

_____ 2. The jigsaw is particularly suited to cutting _____.
   a. metal
   b. curved lines
   c. straight lines
   d. miters

_____ 3. The _____ miter saw is the oldest of the three types of miter saw.
   a. compound
   b. simple
   c. sliding compound-angle
   d. sliding

_____ 4. Drills hold a drill bit, which is gripped by a _____.
   a. shank
   b. miter
   c. chuck
   d. tip

_____ 5. A router _____ holds the router in an inverted position below a table.
   a. plunge
   b. base
   c. fixture
   d. table

21

_____ 6. A _____ is a router-like tool that is smaller and lighter than a router and therefore easier to manipulate.
   a. miter
   b. laminate trimmer
   c. jigsaw
   d. plunger

_____ 7. The _____ sander is the most powerful of the three main types of portable sanders.
   a. pad
   b. disc
   c. palm
   d. belt

_____ 8. Power-driven fastening tools include finish nailers, brad nailers, and _____.
   a. staplers
   b. hammers
   c. jigsaws
   d. laminate trimmers

_____ 9. The smallest standard size that biscuits come in is _____.
   a. 0
   b. 10
   c. 20
   d. 30

_____ 10. Plates are made of compressed _____.
   a. pine
   b. oak
   c. cedar
   d. beech

COMPLETION

Complete each statement.

1. A(n) _____ saw gets its name from its blade, which is shaped like a circle with its cutting teeth arranged around the perimeter.

2. The _____ cuts with a straight blade that reciprocates; that is, it moves up and down.

3. _____ saws, also called _chop saws_ or _drop saws_, are used to make quick, accurate crosscuts.

4. A(n) _____ is an angled cut made along the edge or end of a board.

5. Drill _____ come in a range of sizes and configurations and are designed to bore holes; they consist of a shank, which is inserted into the chuck, and a tip, which does the work.

22

6. The _____ is the primary portable power tool used for shaping wood, though not the only one.

7. Routers may be divided into two main categories: fixed-base routers, also called *standard routers,* and _____ routers.

8. The three main portable sanding tools are _____ sanders, pad sanders, and disc sanders.

9. _____ tools run on compressed air that is usually supplied by an air compressor, which may be gasoline-powered or electric.

10. A(n) _____ joiner, also known as a *biscuit joiner,* is a tool used to join two pieces of material together. It is a great tool for connecting components quickly and is especially useful with manufactured panels of all types.

IDENTIFICATION

Identify each item pictured below. Write the letter of the best answer on the line next to each number.

a.

© Cengage Learning 2014

b.

© Cengage Learning 2014

c.

d.

e.

_____ 1. Routing freehand

_____ 2. Collets and nuts

_____ 3. Saw blades

_____ 4. Biscuits

_____ 5. Jigsaws

SHORT ANSWER

1. What safety habits should you develop, especially with regard to portable power tools?

_____

_____

_____

_____

_____

_____

_____

_____

_____

_____

_____

2. What are some ways you can drill both straight and angled holes accurately?

_____

_____

_____

_____

_____

_____

_____

_____

_____

_____

_____

_____

3. How do you sand a board face with a belt sander?

_____

_____

_____

_____

_____

_____

4. How do you join boards using a plate joiner?

_____

_____

_____

_____

_____

_____

_____

5. What are some ways of storing bits and blades so that they won't be damaged?

_____

_____

_____

_____

_____

_____

_____

CHAPTER

5

Stationary Shop Tools

MULTIPLE CHOICE
Identify the choice that best completes the statement or answers the question.

_____ 1. In a jointer, the _____ provides a surface to support the work, and it is normally set perpendicular to the jointer tables.
   a. infeed
   b. outfeed
   c. fence
   d. cutter head

_____ 2. A _____ is an angled cut made all the way across the edge or end of a board.
   a. crown
   b. bevel
   c. chamfer
   d. miter

_____ 3. The _____ planer, often simply referred to as a *planer*, is a machine used to create boards that are of an even thickness along their whole length.
   a. flattened
   b. stock
   c. push block
   d. thickness

_____ 4. The miter gauge is usually set at _____ degrees to the blade and is used to make square cuts.
   a. 30
   b. 45
   c. 90
   d. 135

_____ 5. _____ is cutting a board in the direction of the grain.
   a. Ripping
   b. Crosscutting
   c. Fencing
   d. Mitering

27

_____ 6. A ____ dado blade, sometimes called an *adjustable dado,* is only one blade, set to oscillate back and forth as it spins, thereby creating a recess.
   a. stack
   b. cutter
   c. chipper
   d. wobble

_____ 7. The ____ may be thought of as a bigger and stronger sibling of the table-mounted router.
   a. shaper
   b. lathe
   c. drill press
   d. grinder

_____ 8. A ____ is a machine tool that uses an abrasive wheel as a cutting device.
   a. lathe
   b. grinder
   c. shaper
   d. router

_____ 9. The degree of coarseness of sandpaper is known as the ____ size.
   a. grit
   b. gram
   c. sand
   d. particle

_____ 10. The ____ is basically a square- or rectangular-shaped recess created to accept the tenon.
   a. rip
   b. shoulder
   c. cheek
   d. mortise

COMPLETION

Complete each statement.

1. The _____ is a machine used to produce a flat surface on a board; it consists of two parallel tables, known as the *infeed* and *outfeed* tables, a moveable fence, and a cutter head.

2. A(n) _____ is an angled cut made partway across the edge or end of a board.

3. In a table saw, the saw blade is mounted on a(n) _____, a metal shaft that is threaded at one end to accept the nut.

4. _____ is an operation in which a cut is made across the gram to change the length of the piece.

5. The _____ angle is the angle of the tooth in relation to the centerline of the blade; it is generally 20 degrees on a rip blade.

28

6. The _____ is an invaluable tool for accurately drilling holes; it is a fixed drill, consisting of a base, column, table, spindle, and drill head.

7. Shapers use _____ rather than bits, as routers do.

8. The _____, the stationary shop tool with the longest history, spins a block of material so that when tools are applied to the block, it is shaped to produce an object that is symmetrical around its axis of rotation.

9. The _____ arm saw consists of a circular saw blade directly driven by an electric motor, which is held in an adjustable yoke that slides along a horizontal arm above a horizontal table surface.

10. A(n) _____ saw is a small electric saw that can cut very intricate patterns and much tighter curves than the band saw.

IDENTIFICATION
Identify each item pictured below. Write the letter of the best answer on the line next to each number.

a.

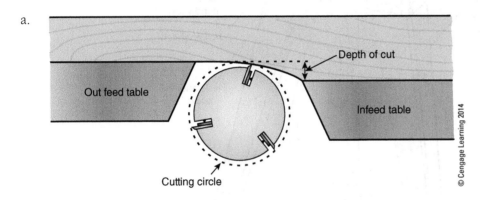

b.

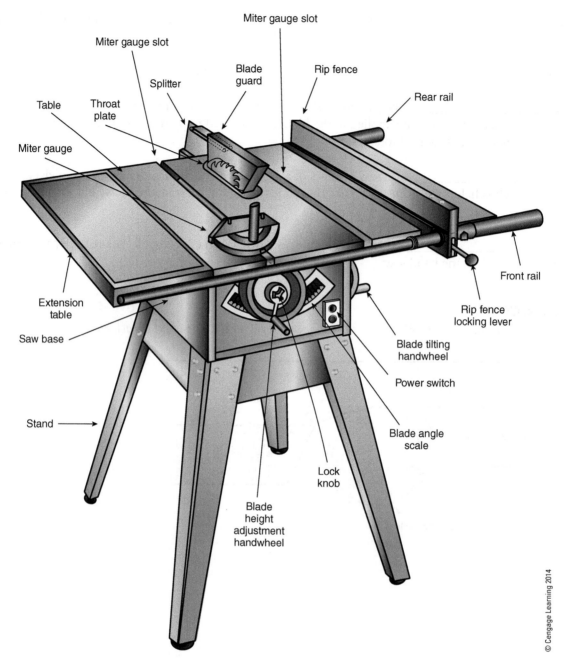

Miter gauge slot

Miter gauge slot

Blade
guard

Rip fence

Splitter

Rear rail

Table

Throat
plate

Miter gauge

Extension
table

Saw base

Front rail

Rip fence
locking lever

Stand

Blade tilting
handwheel

Power switch

Blade angle
scale

Lock
knob

Blade
height
adjustment
handwheel

c.

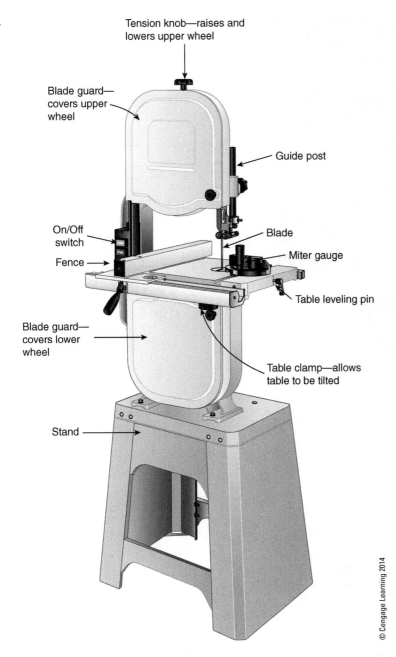

Tension knob—raises and lowers upper wheel

Blade guard—covers upper wheel

Guide post

On/Off switch

Blade

Fence

Miter gauge

Table leveling pin

Blade guard—covers lower wheel

Table clamp—allows table to be tilted

Stand

d.

e.

_____ 1. How a jointer works

_____ 2. External parts of a band saw

_____ 3. External parts of a table saw

_____ 4. Drum sander on a drill press

_____ 5. Three-wing carbide-tipped shaper cutters

SHORT ANSWER

1. What should you do if you are not confident you can perform an operation?

_____

_____

_____

_____

_____

_____

_____

_____

2. How do you edge joint a board?

_____

_____

_____

_____

_____

_____

_____

_____

3. How do you change a band saw blade?

_____

_____

_____

_____

_____

_____

_____

_____

4. How do you use a drill press to drill to an exact depth?

_____

_____

_____

_____

_____

_____

_____

_____

_____

_____

5. How should you treat your machine beds?

_____

_____

_____

_____

_____

_____

_____

_____

_____

_____

CHAPTER

6

Clamps

MULTIPLE CHOICE

Identify the choice that best completes the statement or answers the question.

_____ 1. _____ clamps are similar to bar clamps except that a steel pipe, rather than a bar, holds the jaws.
    a. Pipe
    b. C-
    c. Vise
    d. Web

_____ 2. _____ clamps are metal clamps that are shaped like that letter of the alphabet.
    a. B-
    b. C-
    c. T-
    d. U-

_____ 3. _____ clamps are used primarily used for holding pieces in jigs while they are machined.
    a. Pipe
    b. Bar
    c. Toggle
    d. Web

_____ 4. _____ are sometimes called *wooden parallel clamps*.
    a. Parallel screws
    b. C-clamps
    c. Bench vises
    d. Hand screws

_____ 5. _____ are not used in gluing but are very useful for holding material in place on the work bench.
    a. Bench dogs
    b. Holdfasts
    c. Miter clamps
    d. Frame clamps

_____ 6. Bench vises are used to secure material being worked on at the bench, and they can be used for larger jobs when used in conjunction with _____.
   a. holdfasts
   b. C-clamps
   c. bench dogs
   d. web clamps

_____ 7. Pressure of up to _____ pounds per square foot can be exerted using vacuum clamping.
   a. 300
   b. 600
   c. 900
   d. 1,800

_____ 8. Cam clamps are made up of a locator board, _____, and cams.
   a. pipe
   b. cam dog
   c. holdfast
   d. fence

_____ 9. A veneer _____ is used for attaching veneer to a substrate or to clamp inlaid panels.
   a. press
   b. clamp
   c. hold
   d. vise

_____ 10. The higher the _____ content of wood, the more likely glue stains are to occur.
   a. chlorophyll
   b. iron
   c. tannin
   d. glue

COMPLETION

Complete each statement.

1. _____ blocks are used to make repetitive cuts of the same length and to limit cuts.

2. A(n) _____ is any custom-made accessory that increases the accuracy, speed, or safety of a task.

3. _____ clamps consist of a bar with two jaws; typically, one jaw is connected to a clamp head at one end of the bar and can only be moved a short distance; the other is movable along the whole length of the bar.

4. _____ are metal clamps that are sometimes called _carriage clamps_.

5. _____ clamps are lighter duty than band clamps and use a cloth webbing strap that is an inch wide.

6. _____ clamps operate like a large clothespin.

7. A bench _____ has two jaws, one of which is fixed; the other moves in relation to the first by means of a screw device.

8. _____ clamping is a method of holding material through atmospheric pressure. The air between the work piece and its support is suctioned out using a pump.

9. A(n) _____ is an eccentrically shaped wheel with a pin for placement in a hole in the locator board; in other words, it is not perfectly round.

10. When you use a water-based glue, a reaction occurs between the water in the glue, the _____ in the clamps, and the tannin in the wood to form a stain.

IDENTIFICATION

Identify each item pictured below. Write the letter of the best answer on the line next to each number.

a.

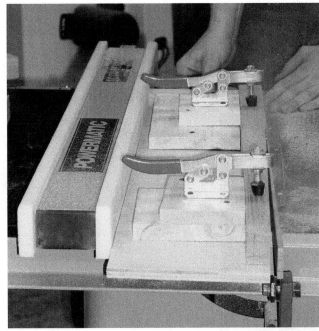

© Cengage Learning 2014

b.

c.

38

d.

e.

_____ 1. Bench vise

_____ 2. Clamps securing a piece to a jig

_____ 3. Bench dog

_____ 4. Web clamp

_____ 5. Stop block

SHORT ANSWER

1. What is the difference between *a feather board* and a *stop block*?

_____

_____

_____

_____

_____

2. What makes pipe clamps frustrating when gluing up a panel?

_____

_____

_____

_____

_____

3. What is the procedure for clamping wide panels?

_____

_____

_____

_____

_____

4. How do miter clamps and hand screws differ?

_____

_____

_____

_____

_____

5. How do you clean up glue squeeze-out?

_____

_____

_____

_____

_____

CHAPTER

7

Fasteners

MULTIPLE CHOICE

Identify the choice that best completes the statement or answers the question.

_____ 1. _____ nails have a thinner cross section and a smaller head than common nails and are used in lighter construction work, such as attaching siding.
  a. Brad
  b. Box
  c. Casing
  d. Finish

_____ 2. _____ are smaller, lighter versions of finish nails.
  a. Brads
  b. Box nails
  c. Casing nails
  d. Common nails

_____ 3. _____ are small brass nails with round heads that are used for decorative purposes or to attach small hardware.
  a. Tacks
  b. Screws
  c. Common nails
  d. Escutcheon pins

_____ 4. When installing screws, a _____ hole, also called a *shank hole,* is drilled in the first piece.
  a. counterbore
  b. plug
  c. clearance
  d. countersink

_____ 5. _____ include panel connectors, cross dowels and bolts, one-piece connectors, dowel screws, corrugated fasteners, and chevrons.
  a. Joint fasteners
  b. Nails
  c. Screws
  d. Nuts

41

_____ 6. Panel connectors are perfect for joining bookcases or entertainment center sections, and they consist of two parts: a _____ and a sleeve nut.
   a. bolt
   b. screw
   c. nail
   d. brad

_____ 7. _____ bolts have a rounded head with a square shoulder.
   a. Insert
   b. Carriage
   c. Hanger
   d. T-

_____ 8. Biscuits, or _____ as they are sometimes called, are football-shaped wafers made of compressed beech.
   a. planes
   b. panels
   c. rafters
   d. plates

_____ 9. Expansion anchors are commonly called molly _____.
   a. screws
   b. bolts
   c. nails
   d. brads

_____ 10. Four types of anchors used to attach objects to poured concrete walls are split-fast anchors, lag shields, lead anchors, and _____ anchors.
   a. steel
   b. gold
   c. aluminum
   d. plastic

COMPLETION

Complete each statement.

1. _____ nails are the stoutest of the nails and are most commonly used in framing.

2. If nails need to be hidden, they are driven close to the surface of the wood, and then set into the wood with a(n) _____, also called a _nail punch._

3. If the screw is to be flush to the surface, the last step before driving the screw is to drill a(n) _____ into which the head of the screw will fit.

4. A(n) _____ is a piece of wire in the shape of a square bracket and might be thought of as a U-shaped nail.

42

5. _____ fasteners are used in the construction of cabinets or furniture that can be assembled or disassembled relatively easily; they are sometimes referred to as *ready-to-assemble (RTA) fasteners.*

6. A(n) _____ is a threaded fastener that has a head at one end and is designed to be inserted through holes in assembled parts and secured with a nut.

7. A(n) _____ is a round wooden pin that fits tightly into corresponding holes to fasten or align to adjacent pieces.

8. One type of anchor is the _____ bolt; it consists of a stove bolt and a spring-loaded toggle and is used to secure objects to hollow walls.

9. Lags _____ are anchors that are used with lag screws.

10. Lead anchors and _____ anchors are also called *inserts;* they are only suitable for hanging light objects.

IDENTIFICATION

Identify each item pictured below. Write the letter of the best answer on the line next to each number.

a.

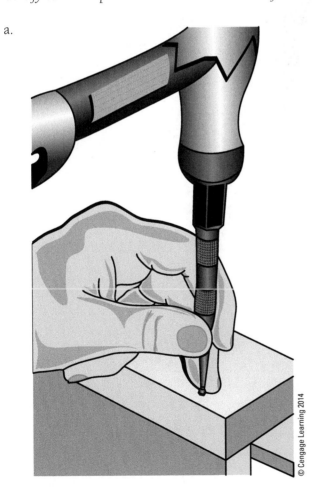

© Cengage Learning 2014

b.

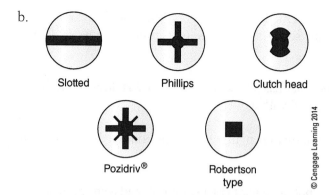

Slotted   Phillips   Clutch head

Pozidriv®   Robertson type

c.

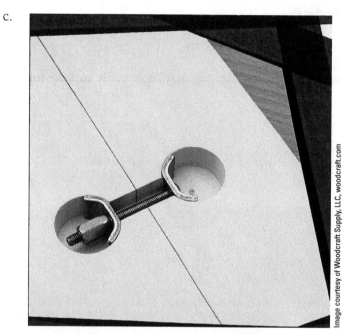

d.

44

e.

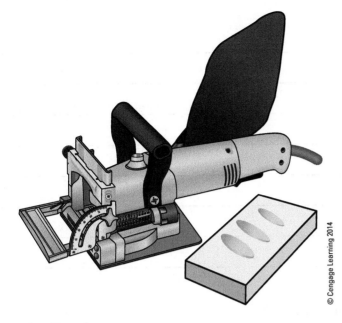

© Cengage Learning 2014

_____ 1. Dowels

_____ 2. Biscuit joiner and biscuits

_____ 3. Using a panel connector

_____ 4. Using a nail set

_____ 5. Screw slots

SHORT ANSWER

1. How do finish nails differ from casing nails?

_____

_____

_____

_____

_____

2. How are screws different from nails? What materials are used to construct them?

_____

_____

_____

_____

_____

3. What is the procedure for driving a nail?

_____

_____

_____

_____

_____

_____

_____

4. How do you install a countersunk screw in hardwood?

_____

_____

_____

_____

_____

_____

_____

5. List the steps involved in making and installing plugs.

_____

_____

_____

_____

_____

_____

_____

CHAPTER

8

Adhesives

MULTIPLE CHOICE

Identify the choice that best completes the statement or answers the question.

_____ 1. _____ adhesion is the bonding that occurs between the adhesive and the wood fibers; tiny fingers of adhesive penetrate the pores and cell cavities of the wood.
   a. Specific
   b. Cohesive
   c. Mechanical
   d. Chemical

_____ 2. _____ resistance is a measure of how the adhesive will stand up to assaults from finishes, cleaning agents, and alcohol.
   a. Water
   b. Heat
   c. Mechanical
   d. Chemical

_____ 3. _____ life describes the amount of time you have for applying the adhesive.
   a. Pot
   b. Shelf
   c. Curing
   d. Set

_____ 4. _____ time also differs from adhesive to adhesive; it is the amount of time it takes for the solvent in the adhesive to evaporate.
   a. Curing
   b. Set
   c. Shelf
   d. Pot

_____ 5. Most adhesives used today are synthetic adhesives, which can be further broken down into two categories, thermoplastic and _____.
   a. natural
   b. hide
   c. PVA
   d. thermosetting

47

_____   6. _____ is a thermoplastic adhesive. It is white in color, so it is often referred to as *white glue;* it dries clear.
   a.   Aliphatic resin
   b.   Polyvinyl acetate
   c.   Epoxy
   d.   Hide glue

_____   7. Plastic resin adhesive is also called _____.
   a.   urea
   b.   alcohol
   c.   ammonia
   d.   white glue

_____   8. _____ is a two-part adhesive; a red, liquid resin is mixed with a tan, powdered catalyst, which acts as a hardener.
   a.   PVA
   b.   Cyanoacrylate
   c.   Resorcinol resin
   d.   Polyurethane

_____   9. There are two basic categories of contact cement, high solvent and _____.
   a.   resorcinol resin
   b.   super glue
   c.   PVA
   d.   neoprene based

_____   10. Hot-melt adhesives and spray adhesives are primarily used as _____ adhesives.
   a.   plastic
   b.   temporary
   c.   permanent
   d.   metal

COMPLETION
Complete each statement.

1. In addition to adhesion, another factor to consider when describing glues is the _____ of the glue, or how well the glue sticks to itself.

2. _____ life, sometimes called *storage time,* is how long the adhesive remains usable.

3. Animal glue, often called _____ *glue,* is manufactured from the hooves, hides, and bones of animals.

4. _____ adhesives were the first adhesives used and have been around for thousands of years; they are made from materials such as bones, blood, hides, eggs, milk, and vegetables.

5. _____ resin is a thermoplastic glue. It is a type of PVA that is yellowish in color and thicker than white glue, so it runs less.

6. Plastic _____ adhesive is also called *urea,* or *urea formaldehyde.*

7. The essential component of _____ glue is diphenylmethane diisocyanate (MDI); it is highly water and heat resistant.

8. _____ is often called *superglue.* Like epoxy cement, it is expensive and only practical for small gluing jobs.

9. _____ cement is different from other adhesives in that it is applied to both surfaces to be bonded, and then each surface is allowed to dry.

10. Hot-melt adhesives will bond many dissimilar types of materials. The adhesive comes in a solid form, usually a stick, and is applied with a special applicator, often called a(n) _____.

IDENTIFICATION

Identify each item pictured below. Write the letter of the best answer on the line next to each number.

a.

© Cengage Learning 2014

b.

c.

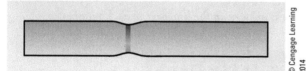

d.

_____ 1. Open assembly

_____ 2. Sunken joint

_____ 3. Closed assembly

_____ 4. Acid brush

SHORT ANSWER

1. What is the difference between *specific* and *mechanical* adhesion?

_____

_____

_____

_____

_____

_____

_____

2. When were synthetic adhesives developed? What major types are available?

_____

_____

_____

_____

_____

_____

_____

_____

3. Give an overview of epoxy glues.

_____

_____

_____

_____

_____

_____

_____

4. What is the basic procedure for using adhesives?

_____

_____

_____

_____

_____

_____

_____

5. Give an overview of contact cements.

_____

_____

_____

_____

_____

_____

_____

52

CHAPTER
9

Wood

MULTIPLE CHOICE
Identify the choice that best completes the statement or answers the question.

____ 1. ____ trees lose their leaves each year.
   a. Deciduous
   b. Coniferous
   c. Softwood
   d. Heartwood

____ 2. The outer part of the cambium is known as the ____.
   a. phloem
   b. xylem
   c. bark
   d. sapwood

____ 3. The ____ of the tree run(s) horizontally across the face of the cross section.
   a. sapwood
   b. heartwood
   c. medullary rays
   d. annual rings

____ 4. Most commercial mills dry their lumber in ____.
   a. saunas
   b. kilns
   c. smokehouses
   d. barns

____ 5. The following describes a ____: a crack caused by a separation of wood fibers along the grain traveling the length of the wood; it is perpendicular to the growth rings.
   a. crack
   b. kink
   c. wane
   d. split

53

_____ 6. The volume of one board foot is _____ cubic inches.
   a. 12
   b. 144
   c. 120
   d. 1,728

_____ 7. A board exhibiting _____ grain on its face was cut along a radius running from the center of the tree outward.
   a. end
   b. radial
   c. longitudinal
   d. edge

_____ 8. The actual moisture in the air is the _____ humidity, and when we talk about the weather being humid, it is this that we are referring to.
   a. specific
   b. mixed
   c. absolute
   d. relative

_____ 9. _____ water accounts for 72 percent of the tree's total moisture content.
   a. Absolute
   b. Free
   c. Bound
   d. Relative

_____ 10. Since most of the wood produced is plainsawn, the majority of the wood we work with will show _____ grain on its faces.
   a. tangential
   b. radial
   c. edge
   d. end

COMPLETION
Complete each statement.

1. It is more accurate to use the term *deciduous* in referring to hardwoods and _____ for softwoods.

2. The very best logs are set aside to be processed into cabinet grade _____.

3. Once the log has been cut into boards, the edges are trimmed off to produce a board of uniform width, and the boards are cut to length. Next, the newly sawn boards go to the _____.

4. A(n) _____ pocket is an opening in the wood containing resin, which may be solid or liquid.

5. Softwood is sold in standard thicknesses, widths, and lengths; it is ordered by its _____ size.

6. A(n) _____ is a unit of volume measurement equivalent to a piece of wood measuring 12 inches wide, 12 inches long, and 1 inch thick.

7. _____ and edge grains are classified as either tangential or radial, depending on how the board was cut in relation to the tree's growth rings.

8. A living tree has a lot of moisture in it, most of it in the form of sap; this is known as _____ water, and it fills the cell cavities of the tree.

9. Because of its unique physical structure, wood expands and contracts in response to changes in the relative _____, which is the ratio of actual moisture in the air to the maximum amount of water the air will hold at its current temperature.

10. Wood is a(n) _____ material, which means that it changes dimension differently in different directions.

IDENTIFICATION

Identify each item pictured below. Write the letter of the best answer on the line next to each number.

a.

© Cengage Learning 2014

b.

Image courtesy of the A. Johnson Co., LLC

c.

© Cengage Learning 2014

d.

© Cengage Learning 2014

56

e.

— Fresh cut
---- After drying

© Cengage Learning 2014

_____ 1. Debarking a log

_____ 2. Hard maple before and after surfacing

_____ 3. Grain of a board from different perspectives

_____ 4. Plainsawn vs. quartersawn lumber

_____ 5. Changes in board shape

SHORT ANSWER

1. What is the significance of the rings in a tree's cross section?

_____

_____

_____

_____

_____

_____

2. What is a warp, and what are the different kinds of warp?

_____

_____

_____

_____

_____

_____

3. Discuss the difference between the two broad categories of softwood lumber: construction and remanufacture.

_____

_____

_____

_____

_____

_____

_____

4. Say a cabinetmaker buys one board that is 4/4 thick, 8 inches wide, and 11 feet long and another that is 8/4 thick, 5 inches wide, and 7 feet long. Calculate the board footage.

_____

_____

_____

_____

_____

_____

_____

_____

5. What are some inexpensive ways to acquire wood?

_____

_____

_____

_____

_____

_____

_____

_____

CHAPTER
10

Panel Products

MULTIPLE CHOICE
Identify the choice that best completes the statement or answers the question.

_____ 1. Plywood sheets are most commonly 4 feet by _____ feet.
   a. 2
   b. 4
   c. 6
   d. 8

_____ 2. Cabinet-grade plywood provides the look of solid wood and is _____.
   a. sturdier
   b. less time-consuming to work
   c. more time-consuming to work
   d. another name for lumber

_____ 3. Cabinet-grade plywood provides the look of solid wood and is _____.
   a. always imported
   b. less expensive to use
   c. more expensive to use
   d. another name for phloem

_____ 4. The highest face grade given to cabinet-grade plywood is _____.
   a. A
   b. B
   c. C
   d. D

_____ 5. The highest back grade given to sheets of plywood is _____.
   a. 1
   b. 2
   c. 3
   d. 4

_____ 6. _____-density fiberboard is a good choice for drawer bottoms and cabinet backs.
  a. Uniform
  b. Low
  c. High
  d. Medium

_____ 7. One drawback of medium-density fiberboard is that it _____.
  a. produces dust when being machined
  b. is too light
  c. must be submerged in water
  d. holds fasteners too tightly

_____ 8. The density of particleboard is _____ the density of hardboard and MDF.
  a. less than
  b. between
  c. the same as
  d. double

_____ 9. _____ is commonly used for "carcass" construction, and is often the major component of inexpensive furniture.
  a. MDF
  b. Hardboard
  c. B-grade plywood
  d. Melamine

_____ 10. The great disadvantage that all panel products share is that the edges of the sheets must be covered; this can be accomplished by _____.
  a. using a router
  b. using a clear lacquer
  c. attaching solid lumber
  d. cross sectioning

COMPLETION

Complete each statement.

1. Cabinet-grade _____ may replace lumber in cabinets or furniture.

2. _____ is created by mixing wood fibers with resin and bonding them together by radio-frequency adhesion or heat.

3. Different grades of cabinet plywood are available. Each sheet has a(n) _____ grade, which appears on the front of the sheet.

4. The three types of veneer cuts are plain sliced, quarter sliced, and _____ cut.

5. _____-density fiberboard is not really used by cabinetmakers; its primary use is in the upholstery industry.

60

6. High-density fiberboard is often referred to as _____.

7. _____ is usually referred to by the acronym MDF.

8. The great disadvantage that all panel products share is that the edges of the sheets must be covered; this can be accomplished using _____ tape.

9. _____ is not as dense as either hardboard or MDF, but it is an adequate and less expensive alternative.

10. _____ is a thermally fused, resin-saturated paper finish applied over a particle-board core.

IDENTIFICATION

Identify each item pictured below. Write the letter of the best answer on the line next to each number.

a.

© Cengage Learning 2014

b.

  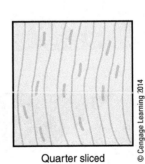

Rotary cut   Plain sliced   Quarter sliced

© Cengage Learning 2014

c.

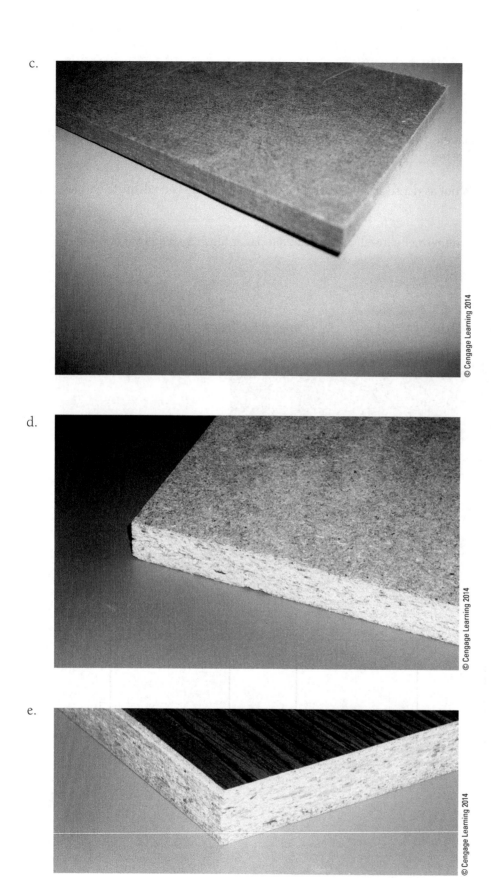

d.

e.

_____ 1. Veneer faces

_____ 2. Melamine

_____ 3. Particleboard

_____ 4. Hardboard

_____ 5. Cabinet-grade plywood

SHORT ANSWER

1. Why is the type of core used in plywood important?

_____

_____

_____

_____

_____

_____

_____

_____

2. When should you use MDF?

_____

_____

_____

_____

_____

_____

_____

_____

3. What are the drawbacks of using MDF?

_____

_____

_____

_____

_____

_____

_____

4. What is the main disadvantage of panel products?

_____

_____

_____

_____

_____

_____

_____

_____

5. How do you apply adhesive-backed edge banding?

_____

_____

_____

_____

_____

_____

_____

_____

CHAPTER

11

Veneer

MULTIPLE CHOICE
Identify the choice that best completes the statement or answers the question.

_____ 1. Stump wood is also known as _____ wood.
  a. butt
  b. crotch
  c. bud
  d. flame

_____ 2. A flame pattern is often exhibited by wood cut from the _____.
  a. butt
  b. crotch
  c. burl
  d. flame

_____ 3. The instrument used in rotary cutting is the _____.
  a. sander
  b. grinder
  c. lathe
  d. router

_____ 4. There are two types of slicing: flat slicing, also called _____ slicing, and quarter slicing.
  a. half
  b. double
  c. whole
  d. plain

_____ 5. In quarter slicing, you end up with _____ flitches.
  a. two
  b. four
  c. eight
  d. ten

_____ 6. Three different patterns may be produced by stay-log cutting: _____, half-round, and back cut.
   a. quilted
   b. rift
   c. full-round
   d. half

_____ 7. Rift cutting is sometimes called _____ cutting, and it results in a very straight-grained veneer.
   a. half-round
   b. comb
   c. burl
   d. flat

_____ 8. It is ideal to have _____ extension cords in a shop.
   a. no
   b. one
   c. three
   d. five

_____ 9. After the veneer has been cut and clipped, it is dried to less than _____ percent moisture content.
   a. 5
   b. 10
   c. 15
   d. 20

_____ 10. The thickest veneers are used as _____.
   a. plies in plywood
   b. peel-and-stick veneers
   c. covering for drywall
   d. edge bands

COMPLETION

Complete each statement.

1. The best logs that are cut, called _____ logs, are sold for veneer production.

2. A(n) _____ is a lump on a tree that is formed by new growth generated to heal an injury.

3. In _____ cutting, the log is turned on a lathe and rotated against a stationary knife.

4. _____ is the method by which most hardwood veneer is cut.

5. In flat slicing, the peeler block is cut in half lengthwise. The two halves are known as _____.

6. _____ cutting is cutting at a 45-degree angle to the annual rings.

7. _____ cuts produce a large U-patterned grain.

8. Veneers 1/28" to 1/40" thick are called _____ veneers.

9. Within a flitch, all of the _____ have a similar grain and color.

10. The thinnest veneers are called _____ veneers.

IDENTIFICATION

Identify each item pictured below. Write the letter of the best answer on the line next to each number.

a.

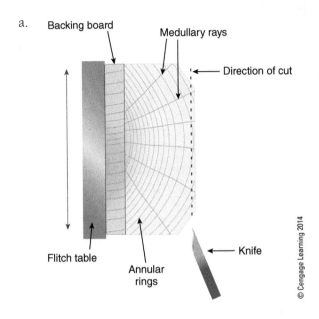

b.

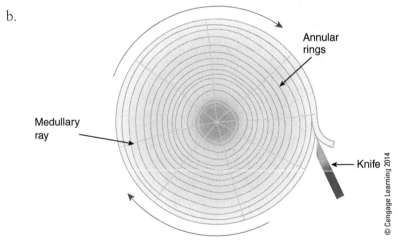

c.

d.

e.

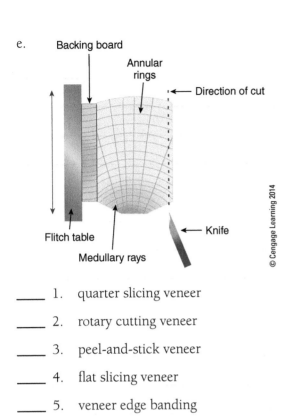

Backing board

Annular rings

Direction of cut

Flitch table

Medullary rays

Knife

© Cengage Learning 2014

_____ 1.   quarter slicing veneer

_____ 2.   rotary cutting veneer

_____ 3.   peel-and-stick veneer

_____ 4.   flat slicing veneer

_____ 5.   veneer edge banding

SHORT ANSWER

1. How are veneer logs processed?

_____

_____

_____

_____

_____

_____

2. How does quarter slicing differ from flat slicing?

_____

_____

_____

_____

_____

_____

3. What is stay-log cutting?

_____

_____

_____

_____

_____

_____

_____

_____

4. Compare and contrast half-round cuts and back cuts.

_____

_____

_____

_____

_____

_____

_____

_____

5. What are the thinnest types of veneer used for?

_____

_____

_____

_____

_____

_____

_____

_____

CHAPTER

12

Synthetic Materials

MULTIPLE CHOICE

Identify the choice that best completes the statement or answers the question.

_____ 1. The core and back of plastic laminates are impregnated with _____ resin.
   a. phenolic
   b. clear melamine
   c. textured
   d. glossy

_____ 2. Rigid laminates are frequently referred to as _____ laminates.
   a. low-temperature
   b. high-pressure
   c. high-temperature
   d. low-pressure

_____ 3. Plastic laminates come in _____ different types
   a. six
   b. five
   c. three
   d. two

_____ 4. Plastic laminate was invented in _____.
   a. 1898
   b. 1900
   c. 1912
   d. 1936

_____ 5. Solid surface material is _____.
   a. only applied to MDF
   b. only applied to plywood
   c. not applied to substrate
   d. applied to the same substrates as plastic laminates

_____ 6. When making solid surface material, the _____ and binder are combined and then cast in a curing process that results in a sheet or shape.
   a. laminate
   b. post-form
   c. filler
   d. vertical surface

_____ 7. The filler used in the production of solid surface is a _____.
   a. gas
   b. synthetic material
   c. liquid crystal
   d. natural mineral

_____ 8. A sheet of solid surface material may be up to 75 percent ATH, which is refined from bauxite ore, a form of _____.
   a. iron
   b. carbon
   c. clay
   d. quartz

_____ 9. The primary disadvantage of solid surface material as compared to plastic laminates is _____.
   a. its higher flammability
   b. its substantially higher cost
   c. that it is easily scratched
   d. its inability to be decorated

_____ 10. Polyester resins, including those that are mixed with acrylic, are used in _____ applications.
   a. high-temperature
   b. low-temperature
   c. high-strength
   d. low-strength

COMPLETION

Complete each statement.

1. Plastic _____ are similar to veneers, but they are synthetic rather than natural.

2. The core and back of plastic laminates are made up of multiple layers of Kraft _____ impregnated with phenolic resin, which is a durable plastic.

3. _____ laminates are designed for surfaces that will have a great deal of use, such as countertops; they are 1/16" thick.

4. _____ laminates are 1/32" thick and are used for the sides and other outside surfaces of cabinets that are subject to less wear than countertops.

5. _____ laminates are no more than 1/32" thick; they are made with flexible resins that allow them to be bent around curved surfaces.

6. _____ surface material is an acrylic material that is manufactured into flat sheets.

7. Most solid surface materials are a combination of two main ingredients: a filler and a(n) _____.

8. The most commonly used filler is _____, or ATH.

9. Two main resins used in the manufacturing process of binders are acrylic and _____.

10. A purely acrylic-based resin yields a sheet that is _____, which means that it can be heated, bent into a new shape, and cooled without any loss of performance characteristics.

IDENTIFICATION

Identify each item pictured below. Write the letter of the best answer on the line next to each number.

a.

© Cengage Learning 2014

b.

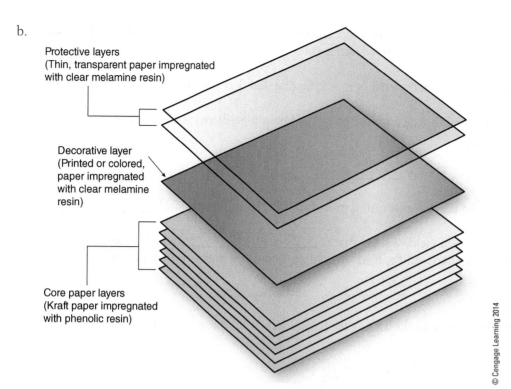

Protective layers
(Thin, transparent paper impregnated
with clear melamine resin)

Decorative layer
(Printed or colored,
paper impregnated
with clear melamine
resin)

Core paper layers
(Kraft paper impregnated
with phenolic resin)

© Cengage Learning 2014

c.

© Cengage Learning 2014

d.

© Cengage Learning 2014

© 2014 Cengage Learning. All Rights Reserved. May not be scanned, copied or duplicated, or posted to a publicly accessible website, in whole or in part.

e.

© Cengage Learning 2014

_____ 1. inlaid solid surface

_____ 2. solid surface materials

_____ 3. plastic laminate composition

_____ 4. post-forming laminate

_____ 5. plastic laminates

SHORT ANSWER

1. How are the layers of plastic laminates bonded together? In what forms are plastic laminates available?

_____

_____

_____

_____

_____

_____

_____

2. What are the major drawbacks of plastic laminates?

_____

_____

_____

_____

_____

_____

_____

3. Where is solid surface material used?

_____

_____

_____

_____

_____

_____

_____

4. What is the difference between acrylic and polyester resins?

_____

_____

_____

_____

_____

_____

_____

5. What are the benefits of using solid surface material instead of plastic laminate?

_____

_____

_____

_____

_____

_____

_____

CHAPTER
13

Construction Methods

MULTIPLE CHOICE

Identify the choice that best completes the statement or answers the question.

_____ 1. The primary disadvantage of solid wood construction is that wood changes in response to seasonal changes in _____.
   a. foliage
   b. atmospheric pressure
   c. humidity
   d. sunlight

_____ 2. _____ was long used to make barrels because it is impervious to water.
   a. White oak
   b. Pine
   c. Beech
   e. Birch

_____ 3. Panel products use _____ solid lumber compared to solid wood construction.
   a. far less
   b. about the same amount of
   c. slightly more
   d. double the

_____ 4. Manufactured panel products include materials such as _____.
   a. melamine
   b. iron
   c. nylon
   d. rayon

_____ 5. In frame-and-panel construction, the _____ can be either solid wood or a manufactured product.
   a. panel
   b. frame
   c. veneer
   d. tape

_____ 6. If you are using a circular saw to cut panel products, the good face of the panel should be _____ to minimize splintering.
   a. removed
   b. down
   c. up
   d. sanded

_____ 7. _____ construction maximizes the advantages of the other two types of construction.
   a. Solid wood
   b. Panel
   c. Veneer
   d. Frame-and-panel

_____ 8. In frame-and-panel construction, the solid wood frame allows for profiles to be _____ on the edges.
   a. milled
   b. carved
   c. glued
   d. taped

_____ 9. A more interesting look can be achieved with frame-and-panel construction, since by its nature it shows _____.
   a. texture
   b. linearity
   c. depth
   d. width

_____ 10. The weight of frame-and-panel construction is _____ that of solid wood construction and manufactured panel construction.
   a. between
   b. less than
   c. about the same as
   d. greater than

COMPLETION

Complete each statement.

1. _____ wood construction is the most ancient of the three building methods.

2. The primary disadvantage of _____ wood construction is that wood is not a static material.

3. Quartersawn _____ is the wood most associated with the Craftsman furniture movement.

4. Manufactured _____ products are materials such as plywood, particleboard, fiberboard, and melamine.

5. _____ is made of thin layers (plies) glued up perpendicular to one another, which diminishes wood movement.

78

6. There are disadvantages to panel construction; one is that the edge of the panel is unattractive and so must be covered with veneer _____.

7. Shelves constructed of plywood, particleboard, or MDF that are longer than 32" should be _____ to prevent sagging under a load.

8. If you are cutting panel products on the table saw, the good side should be face _____ to minimize splintering on the face side of the panel.

9. _____-and-panel construction was developed to deal with the wood movement problem of solid wood.

10. Frame-and-panel construction is _____ in weight than either solid wood construction or manufactured panel construction.

IDENTIFICATION

Identify each item pictured below. Write the letter of the best answer on the line next to each number.

a.

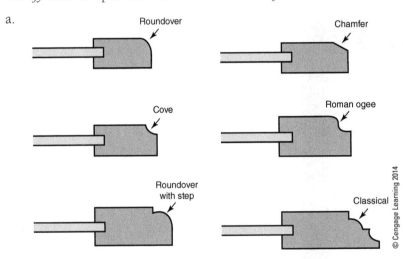

b.

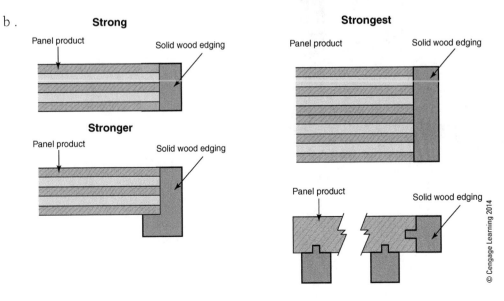

79

c.

d.

e.

_____ 1. frame-and-panel construction

_____ 2. milled profiles

_____ 3. methods of reinforcement

_____ 4. solid wood construction

_____ 5. panel construction

SHORT ANSWER

1. What are the advantages of using solid wood?

_____

_____

_____

_____

_____

_____

2. What are the advantages of using panel construction?

_____

_____

_____

_____

_____

3. What are the disadvantages of using panel construction as opposed to solid wood construction?

_____

_____

_____

_____

_____

4. How can you mill centered grooves in a piece of solid wood?

_____

_____

_____

_____

_____

5. What are the advantages and disadvantages of frame-and-panel construction?

_____

_____

_____

_____

_____

_____

CHAPTER

14

Stock Preparation

MULTIPLE CHOICE

Identify the choice that best completes the statement or answers the question.

____ 1. A ____ is a very important safety device to use when flattening stock on the jointer, and it is easy to make.
   a. jointer
   b. planer
   c. miter saw
   d. push block

____ 2. A rough layout marks out the pieces ____ they will be once they are machined.
   a. smaller than
   b. the same size as
   c. larger than
   d. thicker and shorter than

____ 3. When doing a rough layout, what is a good choice of marking instrument?
   a. chalk
   b. pencil
   c. ballpoint pen
   e. permanent marker

____ 4. When flattening faces, a ____ face should face down if present.
   a. concave
   b. convex
   c. flat
   d. semicircular

____ 5. If possible, you want to joint ____ to avoid tear-out.
   a. perpendicular to the grain
   b. with the crown
   c. with the grain
   d. against the grain

_____ 6. When ripping to width, set the rip _____ to the final width of your work piece.
   a. saw
   b. fence
   c. cut
   d. jointer

_____ 7. The planer maintains the two faces as _____.
   a. acute-angled
   b. parallel
   c. perpendicular
   d. oblique

_____ 8. A _____ cut is a cut made to check the setup of a machine.
   a. miter
   b. rip
   c. check
   d. test

_____ 9. What should you do if the saw you have chosen is not cutting squarely?
   a. Adjust it.
   b. Keep going.
   c. Use a jointer.
   d. Use a planer.

_____ 10. It is necessary to have _____ to make the task at hand, and those to come, easier.
   a. the latest technology
   b. good organization
   c. computers
   d. assistants

COMPLETION

Complete each statement.

1. When selecting stock, it is helpful to do a(n) _____ layout of the components needed.

2. S2S stands for _____.

3. The maximum width piece that you can flatten is determined by the size of your _____.

4. The job of the _____ is to bring material to the desired thickness while maintaining the parallelism of the two faces.

5. The second time the _____ is used is when we need to create an edge that is square to the face of the board.

6. When the board is flat, milled to desired thickness, and has a square edge, it is taken to the _____ saw.

7. The final step in processing stock is to _____ the work piece to finished length.

8. A simple _____ cut and check will determine whether your saw is cutting squarely.

9. The final cut can be made on the radial arm saw, the table saw, or the _____ saw.

10. _____ normally has a straight grain, so it is often used in handles for striking tools, such as hammers, hatchets, and axes.

IDENTIFICATION

Identify each item pictured below. Write the letter of the best answer on the line next to each number.

a.

b.

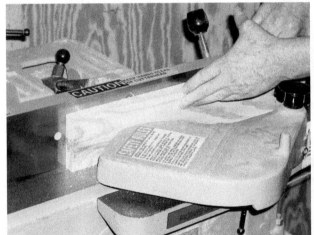

c.

d.

e.

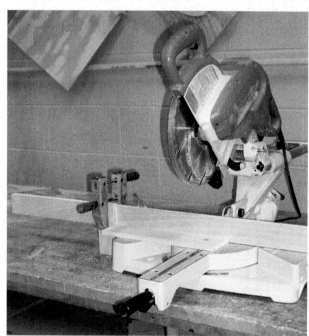

_____ 1. rough layout on stock

_____ 2. squaring an edge

_____ 3. planing to thickness

_____ 4. using a stop block

_____ 5. crosscut to length

SHORT ANSWER

1. When is it essential to flatten a face of the stock?

_____

_____

_____

_____

_____

_____

_____

2. What is the maximum width piece that you can flatten?

_____

_____

_____

_____

_____

_____

_____

_____

3. How do you square an edge?

_____

_____

_____

_____

_____

_____

_____

4. How do you crosscut a piece to finished length?

_____

_____

_____

_____

_____

_____

_____

5. What should you do when you have multiple pieces to process in a project?

_____

_____

_____

_____

_____

_____

_____

CHAPTER
15

Case Joints

MULTIPLE CHOICE
Identify the choice that best completes the statement or answers the question.

_____ 1. In a butt joint, end grain is joined to _____ grain.
   a. face
   b. edge
   c. radial
   d. longitudinal

_____ 2. How many types of rabbet joints are in common use?
   a. one
   b. two
   c. three
   d. four

_____ 3. The pieces in a mitered case joint typically have their ends beveled at a _____-degree angle.
   a. 15
   b. 45
   c. 60
   d. 90

_____ 4. Spline keys are made of _____ material than feather keys.
   a. thinner
   b. stronger
   c. weaker
   d. thicker

_____ 5. The _____ miter joint is an excellent joint because it combines the appearance of a miter joint with the strength of a rabbet-and-dado joint.
   a. feather
   b. lock
   c. spline
   d. key

_____ 6. The tenons in a box joint are known as _____.
   a. notches
   b. boxes
   c. pins
   d. sockets

_____ 7. Prior to the advent of corrugated cardboard boxes, the joint of choice for constructing wooden boxes was the _____ joint.
   a. dovetail
   b. drawer
   c. box
   d. rabbet

_____ 8. The tenons in a dovetail joint are known as _____ and pins.
   a. tails
   b. notches
   c. sockets
   d. doves

_____ 9. There are various types of dovetail joints, but by far the most common are _____ dovetails and half-blind dovetails.
   a. box
   b. edge
   c. through
   d. blind

_____ 10. Jigs can also be built to cut dovetails on the table saw or band saw, but by far the most commonly used method is to use a _____ and some kind of jig.
   a. miter
   b. router
   c. jointer
   d. planer

COMPLETION

Complete each statement.

1. The _____ case joint is the simplest and most basic of all case joints.

2. A(n) _____ is a strip of wood that fits into two matching grooves, one machined in each adjoining piece.

3. One way to reinforce a butt joint is to use _____ blocks, triangular or square pieces of wood used to strengthen and support two adjoining surfaces.

4. A(n) _____ is a two-sided groove along the edge or end of a work piece.

5. A(n) _____ is a three-sided groove.

6. There are three types of keys: spline keys, dovetail keys, and _____ keys.

7. _____ miters are a step up from simple miters.

8. The _____ joint, also called *a finger joint* or *comb joint,* consists of interlocking tenons and notches cut in the ends of adjoining boards.

9. The _____ joint is the quintessential case joint. Some consider the hand-cut version the holy grail of woodworking.

10. In a dovetail joint, both the tails and pins fit into the recesses on the mating pieces. These recesses are called _____.

IDENTIFICATION

Identify each item pictured below. Write the letter of the best answer on the line next to each number.

a.

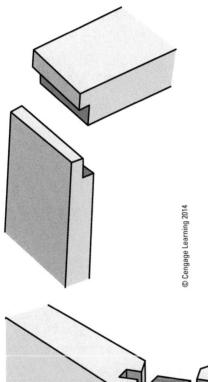

© Cengage Learning 2014

b.

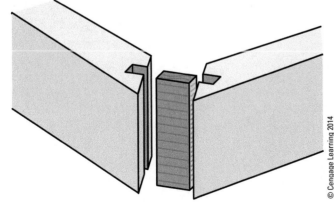

© Cengage Learning 2014

c.

d.

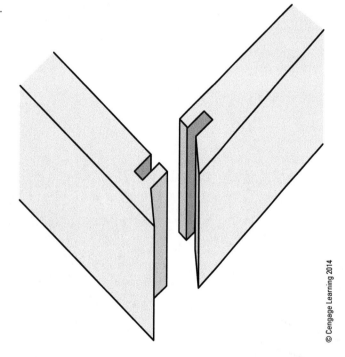

92

e.

_____ 1. lock miter joint

_____ 2. double-rabbet joint

_____ 3. splined miter joint

_____ 4. half-blind dovetail jig

_____ 5. box joint

SHORT ANSWER

1. How do you make a rabbet joint on a table saw?

_____

_____

_____

_____

_____

_____

2. How do you machine a dado-and-rabbet joint?

_____

_____

_____

_____

_____

_____

3. How do you make a splined miter joint?

_____

_____

_____

_____

_____

_____

_____

4. How do you choose the size and spacing of fingers in a box joint?

_____

_____

_____

_____

_____

_____

_____

5. Give a brief procedure for cutting a dovetail joint by hand.

_____

_____

_____

_____

_____

_____

_____

_____

CHAPTER
16

Frame Joints

MULTIPLE CHOICE

Identify the choice that best completes the statement or answers the question.

_____ 1. Frames have _____ main uses in cabinet construction.
   a. two
   b. three
   c. four
   d. five

_____ 2. In a mitered frame joint, each member is cut at a _____-degree angle.
   a. 30
   b. 45
   c. 75
   d. 90

_____ 3. A pocket-hole joint uses _____ as reinforcements.
   a. screws
   b. nails
   c. biscuits
   d. dowels

_____ 4. Miter joints, like butt joints, are _____.
   a. easy to assemble
   b. unable to be reinforced
   c. structurally weak
   d. structurally strong

_____ 5. Lap joints may meet to form an L, a T, or a(n) _____.
   a. A
   b. C
   c. V
   d. X

_____   6. _____, also known as *canoe wood,* has a straight grain and fine texture; it is often used for furniture and cabinetwork, carving, and musical instruments.
    a. Pine
    b. Oak
    c. Ash
    d. Poplar

_____   7. A joint in which the end of one piece is joined to the end of another piece is called a _____ joint.
    a. miter
    b. slip
    c. bridle
    d. rein

_____   8. A blind mortise and tenon is also called a _____ mortise and tenon.
    a. stopped
    b. jigged
    c. loose
    d. haunched

_____   9. An invisible wedged mortise and tenon is also called a _____-wedged mortise and tenon.
    a. visible
    b. haunched
    c. fox
    d. through

_____   10. The pegged mortise-and-tenon will still work if _____.
    a. the peg snaps off
    b. only one piece of wood is used
    c. the glue fails
    d. the peg disintegrates

COMPLETION

Complete each statement.

1. _____ frames are installed inside a case, tying the case together and supporting and separating drawers.

2. _____ frame joints form a neat right-angle corner, and if they fit well, the seam is almost indiscernible.

3. The _____-and-stub tenon joint is a fairly light-duty joint, appropriate for building frames that will be anchored to a case, such as face frames.

4. The _____-and-stick joint is a more elegant version of a related tenon joint, and a stronger one, too.

5. A slip joint in which the end of one piece is joined to the middle of another piece is called a(n) _____ joint.

6. The _____-and-tenon joint is woodworking's essential frame joint, in the same way that the dovetail joint is the ultimate case joint.

7. Tenons are easily made on the table saw with a tenoning _____.

8. A(n) _____ mortise-and-tenon joint has a tongue that projects from the tenon's shoulder between the edge of the tenon and the edge of the rail.

9. There are two types of _____ mortise-and-tenon joints: visible and hidden.

10. The pegged mortise-and-tenon is usually a blind mortise and tenon that is further reinforced with _____.

IDENTIFICATION

Identify each item pictured below. Write the letter of the best answer on the line next to each number.

a.

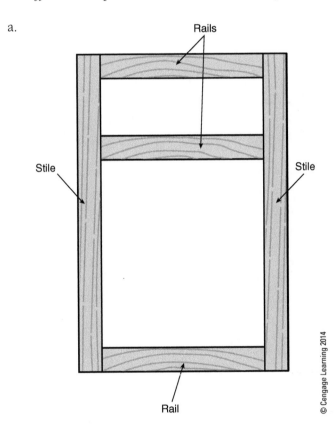

Rails

Stile

Stile

Rail

© Cengage Learning 2014

b.

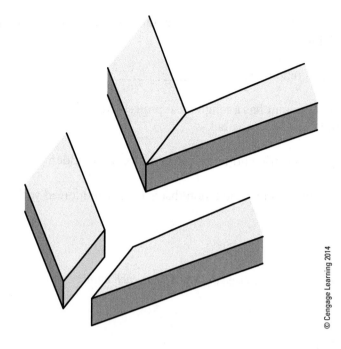

c.

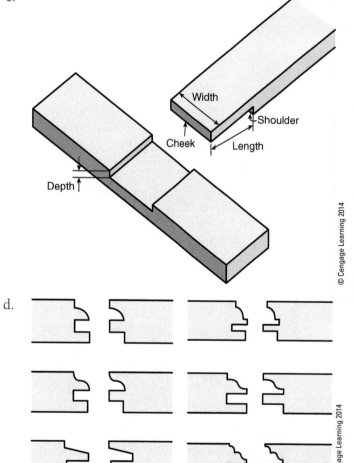

Width

Shoulder

Cheek

Length

Depth

d.

98

e.

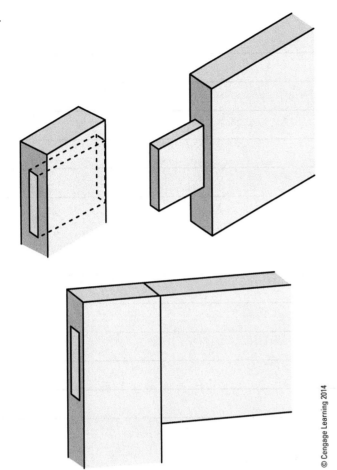

_____ 1. face frame

_____ 2. half-lap joint

_____ 3. miter joint

_____ 4. cope-and-stick profiles

_____ 5. through mortise and tenon

SHORT ANSWER

1. How do you create a pocket joint?

_____

_____

_____

_____

_____

_____

2. How do you machine a half-lap joint?

_____

_____

_____

_____

_____

3. How do you make a mitered half-lap joint?

_____

_____

_____

_____

_____

4. How do you make a groove-and-stub tenon joint?

_____

_____

_____

_____

_____

5. How do you make the mortise in a blind mortise-and-tenon joint?

_____

_____

_____

_____

_____

100

CHAPTER
17

Rail Joints

MULTIPLE CHOICE

Identify the choice that best completes the statement or answers the question.

_____ 1. The _____-tenon joint is a very functional joint in furniture that needs to be disassembled occasionally; it is a knockdown joint commonly used in trestle tables and beds.
   a. mitered
   b. multiple
   c. twin
   d. tusk

_____ 2. In a _____-tenon joint, the tenon itself has a mortise through it, which allows the insertion of a removable wedge that locks the joint together.
   a. mitered
   b. multiple
   c. twin
   d. tusk

_____ 3. If you are making a through cut on a table saw, you can still use the fence as a stop, provided you attach a _____ to it.
   e. half fence
   f. multiple fence
   g. twin-tenon joint
   h. test piece

_____ 4. The bridle joint is most often seen in rail construction for tables that have _____ aprons.
   a. straight
   b. curved
   c. no
   d. twin

_____ 5. The _____ joint used in rail joinery is not a particularly attractive joint, so it is most often used on heavy-duty post-and-rail structures, such as workbenches. Although not especially good looking, it is a very sturdy joint.
   a. mortise-and-tenon
   b. slip
   c. lap
   d. dovetail

_____ 6. The full-lap joint has material removed from _____ member(s).
   a. no
   b. one
   c. two
   d. a varying number of

_____ 7. The dovetail joint _____.
   a. snaps easy
   b. strongly resists forces placed on it
   c. uses a tenon extending through a mortise
   d. has a socket cut into the rail

_____ 8. Dowel joints are often strengthened using _____.
   a. wedges
   b. bed bolts
   c. corner braces
   d. mortises

_____ 9. A metal corner plate allows aprons to be assembled to legs with _____ joints.
   a. lap
   b. butt
   c. mortise-and-tenon
   d. dovetail

_____ 10. The bolt head in a bed bolt may be recessed and covered by a _____.
   a. decorative plate
   b. wedge
   c. tenon
   d. corner plate

COMPLETION

Complete each statement.

1. _____-tenon joints are used to join a drawer rail to a leg.

2. The _____-tenon joint, consisting of two or more tenons and corresponding mortises, is used on wide rails to counteract the expansion and contraction that would occur with a single-wide tenon.

3. A(n) _____ tenon extends through its mortise and beyond.

4. Slip joints are mostly used in frame construction, but one form of the slip joint, the _____ joint, is very good in rail joinery as well.

5. The _____-lap joint only has material removed from one member.

6. The _____-lap joint has a notch cut in the edge of each of the adjoining members, allowing them to interlock.

7. When crosscutting on the table saw, you should never use the rip _____ as a stop when making a through cut.

8. The _____ joint can be used for joining of a top drawer rail to a table legs.

9. The _____ joint is not a particularly good rail joint and is only suitable for small, light-duty constructions, unless it is reinforced.

10. Bed _____ are instances of specialized connectors used in making rail joints for beds.

IDENTIFICATION

Identify each item pictured below. Write the letter of the best answer on the line next to each number.

a.

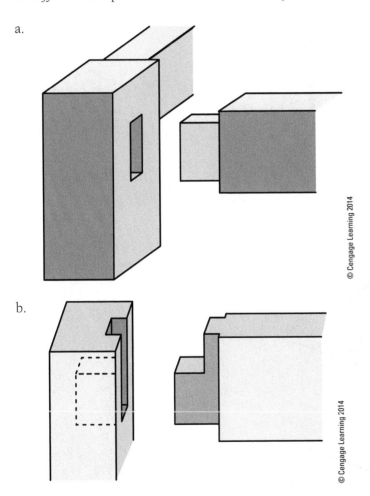

© Cengage Learning 2014

b.

© Cengage Learning 2014

c.

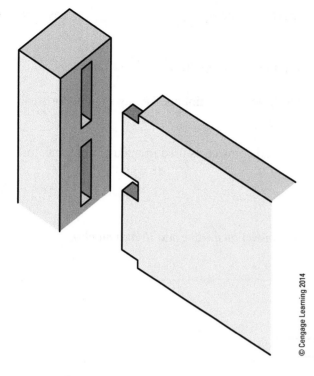

d.

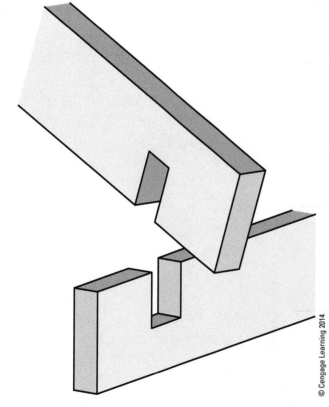

e.

Leg

Rail

© Cengage Learning 2014

____ 1. dovetail joint

____ 2. haunched tenon

____ 3. edge-lap joint

____ 4. multiple-tenon joint

____ 5. mitered tenon

SHORT ANSWER

1. Where are slip joints used in rail joinery?

_____

_____

_____

_____

_____

_____

_____

2. What is the difference between the two lap joints used in rail joinery?

_____

_____

_____

_____

_____

_____

_____

_____

3. What are the characteristics of beechwood?

_____

_____

_____

_____

_____

_____

_____

4. How does a metal corner plate work?

_____

_____

_____

_____

_____

_____

_____

5. How do bolt and barrel nut assemblies work?

_____

_____

_____

_____

_____

_____

_____

CHAPTER
18

Housed Joints

MULTIPLE CHOICE
Identify the choice that best completes the statement or answers the question.

_____ 1. The butt joint is _____.
   a. a high-end joint
   b. mainly used for aesthetic purposes
   c. mainly used for utilitarian purposes
   d. a variation on the dado-and-rabbet joint

_____ 2. The _____-dado housing has a dado that stops short of the front of the case but runs fully to the back.
   a. stopped
   b. rabbet
   c. blind
   d. through

_____ 3. Both the stopped-dado and _____-dado housings require that the shelf be trimmed to fit the dado.
   a. through
   b. dovetail
   c. rabbet
   d. blind

_____ 4. A simple, shop-built _____ makes routing dadoes quick and easy, especially on large sheet goods that might be tricky to maneuver on the table saw.
   a. laminate
   b. stop block
   c. jig
   d. router

_____ 5. In the dado-and-rabbet joint, the dado is milled _____.
   a. into the case
   b. into the shelf
   c. into the divider
   d. from the rabbet

_____ 6. A shelf housed in a case can hold more weight if the rabbet is milled on its _____ surface.
   a. bottom
   b. top
   c. front
   d. back

_____ 7. In a tongue-and-dado housing, the tongue has _____ on both sides of it.
   a. rabbets
   b. dadoes
   c. blinds
   d. shoulders

_____ 8. In a dado-and-spline joint, the spline should be set into the case about _____ of the thickness of the case material.
   a. a quarter
   b. a third
   c. half
   d. two-thirds

_____ 9. The sliding dovetail joint has _____ mechanical strength than the simple dado housing.
   a. much less
   b. slightly less
   c. slightly more
   d. much more

_____ 10. The multiple-tenon housing consists of multiple tenons on the shelf or case divider that fit into corresponding _____ in the case.
   a. mortises
   b. dadoes
   c. splines
   d. rabbets

COMPLETION
Complete each statement.

1. The _____ joint is the most commonly used housing joint in cabinet construction.

2. A dado that runs from one end of the case to the other is called a(n) _____ dado.

3. A butt joint that is used as a housing joint must be _____.

4. The _____-dado housing is stopped short of both the front and the back of the case.

5. The dado-and-rabbet housing joint has a(n) _____ on the end of the shelf or divider.

6. The _____-and-dado housing is similar to the dado and rabbet, but the rabbet is replaced by another structure.

7. The dado-and-_____ joint is a good choice when working with MDF or particle-board.

8. The _____ joint is a more elaborate version of the dado joint.

9. A slightly altered version of the sliding dovetail, with only one sloping side, is the sliding _____-dovetail.

10. The multiple-tenon housing is a(n) _____-and-tenon joint.

IDENTIFICATION

Identify each item pictured below. Write the letter of the best answer on the line next to each number.

a.

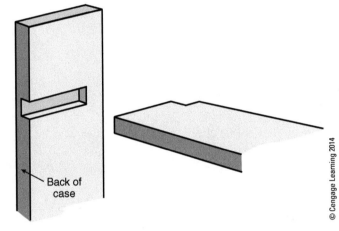

Back of case

© Cengage Learning 2014

b.

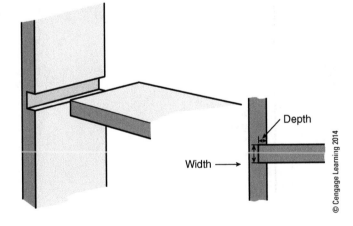

Depth

Width

© Cengage Learning 2014

c.

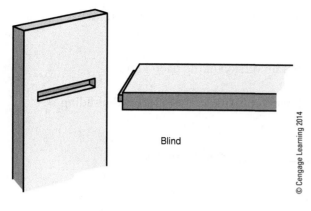

Blind

d.

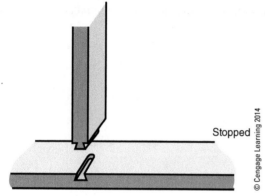

Stopped

e.

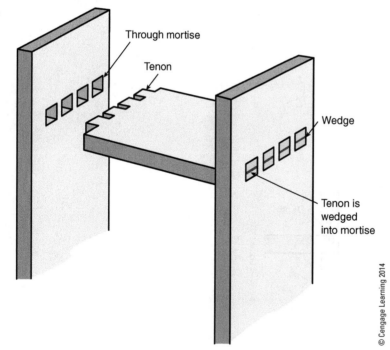

Through mortise

Tenon

Wedge

Tenon is
wedged
into mortise

_____ 1. sliding-dovetail joint

_____ 2. multiple-tenon housing

_____ 3. dado housing

_____ 4. stopped-dado joint

_____ 5. tongue-and-dado joint

SHORT ANSWER

1. What reinforcements are used in the construction of butt joints?

_____

_____

_____

_____

_____

_____

_____

2. What is one disadvantage of dado joints, and how can a cabinetmaker avoid this problem?

_____

_____

_____

_____

_____

_____

_____

3. How do you make a jig for routing dadoes?

_____

_____

_____

_____

_____

_____

_____

4. What is soft maple used for?

_____

_____

_____

_____

_____

_____

_____

_____

5. Describe how to make a dado-and-rabbet housing.

_____

_____

_____

_____

_____

_____

_____

_____

_____

CHAPTER
19

Making Wide Panels, Thick Blanks, and Corner Joints

MULTIPLE CHOICE
Identify the choice that best completes the statement or answers the question.

_____ 1. Reinforced joints for wide panels include those that use _____.
  a. dowels
  b. glue
  c. miters
  d. fingers

_____ 2. The use of butterfly keys is a traditional element of _____ joinery.
  a. Italian
  b. Japanese
  c. Indian
  d. Mexican

_____ 3. In a tongue-and-groove joint, the tongue thickness is typically _____ the thickness of the stock being used.
  a. a quarter
  b. a third
  c. half
  d. three-fifths

_____ 4. The shiplap joint is _____ than the tongue-and-groove joint.
  a. more slowly cut
  b. a better joint
  c. more closely fitting
  d. more quickly cut

_____ 5. Clamps should be spaced _____ inches apart, alternating on the bottom and top of the panel.
  a. 2 to 4
  b. 4 to 8
  c. 8 to 12
  d. 12 to 14

113

6. When making a thick blank, the individual pieces must have ____ faces.
   a. large
   b. small
   c. parallel
   d. perpendicular

7. Corner joints are created by attaching the ____ of one piece to the ____ of another.
   a. face, face
   b. edge, end
   c. edge, edge
   d. edge, face

8. The advantage of the rabbet-and-groove joint is that ____.
   a. it has a rabbet on both pieces
   b. it has a prominent seam
   c. it must use a reveal
   d. it locks in place

9. The edge-miter joint looks similar to the case-miter joint, but differs in its ____.
   a. exclusive use of a lock
   b. grain orientation
   c. inability to conceal gram change
   d. easy assembly

10. The lock-miter joint is most frequently used to make ____.
    a. edge-miter joints
    b. case-miter joints
    c. grooves
    d. tongues

COMPLETION

Complete each statement.

1. _____ panels are manufactured by joining two or more narrow boards together edge to edge.

2. The use of _____ keys consists of insetting a double-dovetail shaped piece into the main pieces to help lock them together.

3. _____ joints used for joining boards edge to edge for wider panels include tongue-and-groove, shiplap, glue, finger, and lock-miter joints.

4. The _____ joint has rabbets cut into the opposite faces of adjoining boards. It cannot keep surfaces flush.

5. _____ blanks are created when pieces are glued up face to face.

6. _____ joints are constructed by joining the edge of one board to the face of another.

7. Several types of rabbet joints may also be employed in forming corner joints; they are the single-rabbet, double-rabbet, and rabbet-and-_____ joints.

8. One option with the single-rabbet joint is to cut the rabbet slightly less than the thickness of the mating piece, creating a(n) _____, which allows part of the piece being covered to show.

9. In the _____-miter joint, the end of one piece is joined to the end of another.

10. With the _____-miter joint, the edge of one piece is joined to the edge of its mate.

IDENTIFICATION

Identify each item pictured below. Write the letter of the best answer on the line next to each number.

a.

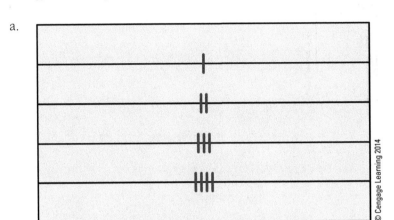

© Cengage Learning 2014

b.

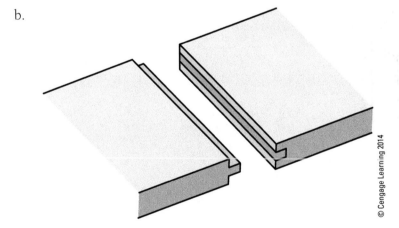

© Cengage Learning 2014

c.

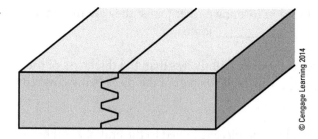

d.

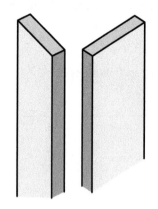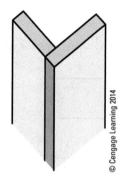

e.

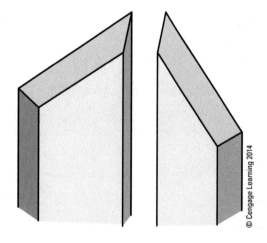

_____ 1. edge-miter joint

_____ 2. corner joint

_____ 3. finger joint

_____ 4. tongue-and-groove joint

_____ 5. panel marked for easy reassembly

SHORT ANSWER

1. How do you minimize warping in wide panels?

_____

_____

_____

_____

_____

_____

2. How are butterfly keys constructed and used?

_____

_____

_____

_____

_____

_____

3. How do you mill a tongue-and-groove joint for a wide panel?

_____

_____

_____

_____

_____

_____

4. How are thick blanks created?

_____

_____

_____

_____

_____

_____

_____

5. What are the safety precautions you should take when working with a shaper?

_____

_____

_____

_____

_____

_____

_____

CHAPTER
20

Cabinets

MULTIPLE CHOICE

Identify the choice that best completes the statement or answers the question.

_____ 1. Frameless cabinets were initially introduced in _____.
    a. Europe
    b. New York
    c. Japan
    d. China

_____ 2. In a cabinet with a face frame, the bottom is usually recessed into a _____.
    a. frame
    b. top
    c. dado
    d. rabbet

_____ 3. In a face frame, the _____ fit between the stiles.
    a. mullions
    b. cross stiles
    c. rungs
    d. rails

_____ 4. In European cabinets, a series of _____-mm holes are drilled into the sides.
    a. 5
    b. 12
    c. 20
    d. 32

_____ 5. European-style cabinets _____.
    a. use the extended sides as the foundation
    b. are usually set on separate feet
    c. use 32-mm pegs as feet
    d. are set on a pedestal

_____ 6. European cabinets are _____.
   a. rarely found
   b. built with frames
   c. produced rapidly
   d. slowly produced

_____ 7. The standard height of kitchen base cabinets, including the countertop, is _____ inches.
   a. 20
   b. 32
   c. 36
   d. 48

_____ 8. Kitchen base cabinets are _____ inches deep, excluding the countertop.
   a. 12
   b. 20
   c. 24
   d. 36

_____ 9. The countertop in bathroom vanities is generally _____ inch(es) in thickness.
   a. a half
   b. one
   c. one and a quarter
   d. one and a half

_____ 10. Auxiliary fences are also known as _____ fences.
   a. sacrificial
   b. dado
   c. rabbet
   d. push

COMPLETION

Complete each statement.

1. Cabinets without face frames are called _frameless_ or _____ cabinets.

2. The essential parts of a cabinet are the sides, the bottom, and the _____.

3. Cabinets with face frames have vertical members known as _____.

4. A face frame may be divided by a vertical member, called a(n) _____, which fits between the rails.

5. The European cabinet is also known as the _____ millimeter cabinet.

6. Instead of the tops and bottoms of frameless cabinets being set into _____, as they typically are in face-frame construction, they are usually doweled or screwed into place through the sides.

7. European cabinetry is _____ (less/more) expensive than traditional face-frame cabinetry.

8. Kitchen _____ cabinets are 34 ½" high; they sit on the floor.

9. Most kitchen _____ cabinets are 12" deep and are usually 30" high, although it is possible to purchase them in other dimensions.

10. Bathroom _____, which are cabinets that contain a sink, are the same depth as kitchen base cabinets but are a total height of 34" with the countertop.

IDENTIFICATION

Identify each item pictured below. Write the letter of the best answer on the line next to each number.

a.

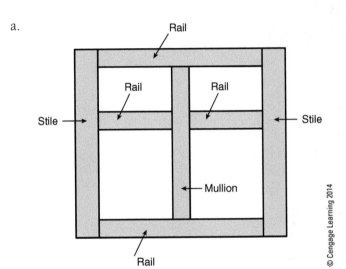

b.

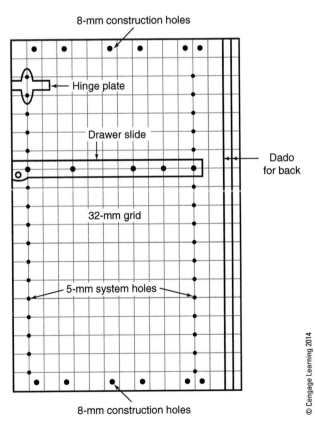

c.

d.

Top view

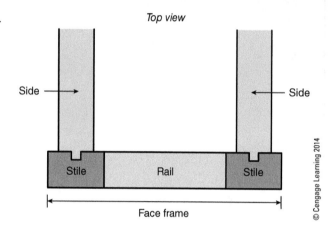

Side ———→   ←——— Side

| Stile | Rail | Stile |

Face frame

e.

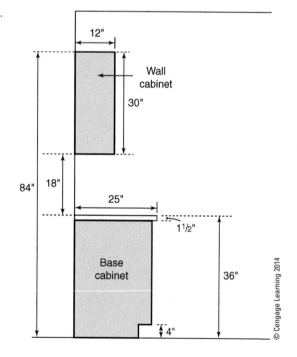

12"

Wall cabinet

30"

84"

18"

25"

1½"

Base cabinet

36"

4"

122

_____ 1. veneer tape

_____ 2. side of 32-mm cabinet

_____ 3. face frame components

_____ 4. face frame to cabinet attachment: tongue-and-groove joinery

_____ 5. kitchen cabinet dimensions

SHORT ANSWER

1. What are the two main types of cabinets? How are they similar?

_____

_____

_____

_____

_____

_____

_____

_____

_____

2. How do you build a face frame using half-lap joints?

_____

_____

_____

_____

_____

_____

_____

_____

_____

_____

3. What has caused the upsurge in the popularity of European cabinets?

_____

_____

_____

_____

_____

_____

4. What kinds of alterations exist to the dimensions of kitchen wall cabinets?

_____

_____

_____

_____

_____

_____

5. What is an auxiliary fence?

_____

_____

_____

_____

_____

_____

_____

CHAPTER
21

Cabinet Doors and Drawers

MULTIPLE CHOICE
Identify the choice that best completes the statement or answers the question.

_____ 1. The simplest type of door has _____ frame.
- a. a flat-panel
- b. a raised-panel
- c. no
- d. a narrow

_____ 2. A solid wood panel should never be _____.
- a. glued into its frame
- b. given room to account for expansion
- c. decorated with a profile on their edges
- d. raised

_____ 3. The most time-consuming way to connect door stiles and rails is with _____ joints.
- a. cope-and-stick
- b. raised-panel
- c. biscuit
- d. mortise-and-tenon

_____ 4. When making lap joints, _____ of the material's thickness is removed from the back of the stile and the front of the rail.
- a. one-quarter
- b. one-third
- c. two-fifths
- d. one-half

_____ 5. The dowel joint, biscuit joint, and pocket-screw method of joining the stiles and rails are all simple _____ joints between the stiles and rails.
- a. dovetail
- b. butt
- c. mortise
- d. lap

_____ 6. A flat door should have its handle or knob set _____ inch(es) in from the edge of the door and 4 inches from the corner.
   a. 1
   b. 2
   c. 3
   d. 4

_____ 7. _____ doors sit partly into the face frame or cabinet, if it is without a face frame.
   a. Overlay
   b. Flush
   c. Inlay
   d. Drawer

_____ 8. One option to deal with double doors is to add a(n) _____ in the center of the frame between the two rails.
   a. astragal
   b. raised panel
   c. stile
   d. mullion

_____ 9. A drawer lock is similar to a _____ joint but requires a special drawer-lock bit used with a table-mounted router.
   a. biscuit
   b. dowel
   c. dovetail
   d. rabbet-and-dado

_____ 10. A sliding dovetail is also called a _____ dovetail.
   a. dado
   b. half-blind
   c. through
   d. rabbet

COMPLETION

Complete each statement.

1. The _____ door is a style of door with medium complexity that has a frame consisting of stiles and rails.

2. The style of door that is the most complicated and time consuming to build is the _____ door.

3. In a raised-panel door, the width of the panel must be _____ than the distance from the bottom of the groove in one stile to the bottom of the groove in the other stile.

4. The strongest way to connect the door stiles and rails is with _____ joints.

126

5. The groove-and-_____ tenon method of joining stiles and rails is closely related to the cope-and-stick joint, but it can be machined using just the table saw.

6. Doors are built to be 3/4" wider and taller than the door opening, yielding a 3/8" overlap all the way around.

7. A(n) _____ is a strip of wood attached to either the front or the back of one door to eliminate the gap between the two.

8. Like doors, drawers may be _____, overlay, or flush.

9. The _____ dovetail is the one most often seen in drawer construction.

10. If the drawer is very wide, a centered wooden strip, called a(n) _____, can be installed to prevent sagging.

IDENTIFICATION

Identify each item pictured below. Write the letter of the best answer on the line next to each number.

a.

© Cengage Learning 2014

b.

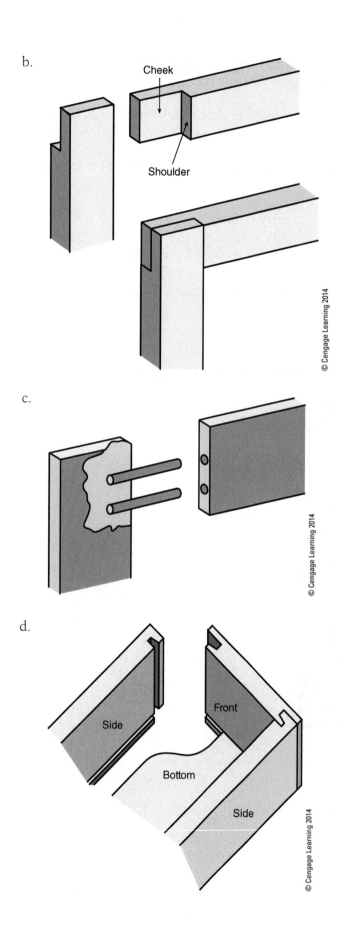

Cheek

Shoulder

c.

d.

Side

Front

Bottom

Side

e.

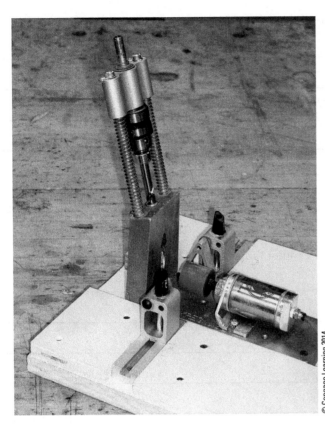

© Cengage Learning 2014

_____ 1. routed drawer lock joint

_____ 2. dowel joint

_____ 3. lap joint

_____ 4. pocket-hole jig

_____ 5. flat-panel doors

SHORT ANSWER

1. How do you make a raised panel on a router table?

_____

_____

_____

_____

_____

_____

_____

2. How do you mill a groove-and-stub tenon joint on a router table?

_____

_____

_____

_____

_____

_____

3. How do you mill a rabbet-and-dado joint on a router table?

_____

_____

_____

_____

_____

4. How do you machine a sliding dovetail?

_____

_____

_____

_____

_____

5. Describe the attributes of cherry wood.

_____

_____

_____

_____

_____

CHAPTER
22

Tables and Desks

MULTIPLE CHOICE

Identify the choice that best completes the statement or answers the question.

____ 1. Aprons can be rectangular or ____.
   a. square
   b. tapered
   c. relieved
   d. sculpted

____ 2. Bedside tables are often called ____.
   a. night stands
   b. coffee tables
   c. sideboards
   d. hall tables

____ 3. The basic type of table is sometimes called a ____ table.
   a. trestle
   b. pedestal
   c. dining
   d. leg-and-apron

____ 4. Tabletop blocks, used to attach the tabletop to the frame, are also known as ____ blocks.
   a. figure-eight
   b. trestle
   c. button
   d. mortise

____ 5. Pedestal tables consist of a central ____ that supports the top.
   a. trestle
   b. pillar
   c. leaf
   d. apron

131

_____ 6. What is the bevel angle for a 12-sided pedestal?
   a. 15 degrees
   b. 30 degrees
   c. 90 degrees
   d. 360 degrees

_____ 7. _____ is defined as the distance from the floor to the underside of the desktop.
   a. Knee room
   b. Leg room
   c. Thigh room
   d. Apron clearance

_____ 8. The _____ desk is also known as a _plantation desk._
   a. postmaster's
   b. schoolmaster's
   c. computer
   d. writing

_____ 9. A _____ is a combination desk and bookcase.
   a. plantation desk
   b. secretary
   c. schoolmaster's desk
   d. writing desk

_____ 10. The sight angle of a computer desk should not exceed _____ degrees.
   a. 30
   b. 45
   c. 60
   d. 90

COMPLETION

Complete each statement.

1. _____ are horizontal pieces that run between the legs and support the tabletop.

2. _____ tables generally sit in front of a couch and are sometimes called _cocktail tables._

3. Side tables are sometimes called _____; these are oblong tables originally designed to be set against a wall close to the kitchen.

4. There are two commonly used types of tabletop fasteners, both of which are metal; one is shaped like the figure eight, and the other is called a(n) _____-clip.

5. Not all tables are leg-and-apron tables. There are also _____ tables and pedestal tables.

6. A(n) _____ table is any table that has leaves hinged to the tabletop that hang vertically when not in use.

7. The slant-top desk is also known as a(n) _____ desk or a _stand-up desk._

132

8. A modesty or _____ panel often closes the kneehole on the far side of the knee-hole desk.

9. A(n) _____ desk, also called a *tambour desk*, is characterized by a tambour curtain that pulls down to completely enclose the writing surface.

10. The sight angle of a computer desk is defined as the angle between the sight line to the _____ and the sight line to the monitor.

IDENTIFICATION

Identify each item pictured below. Write the letter of the best answer on the line next to each number.

a.

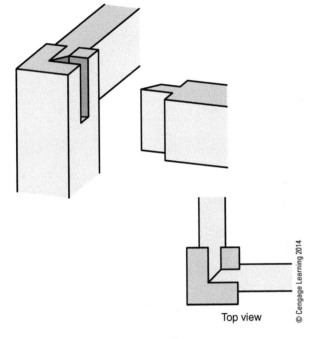

Top view

© Cengage Learning 2014

b.

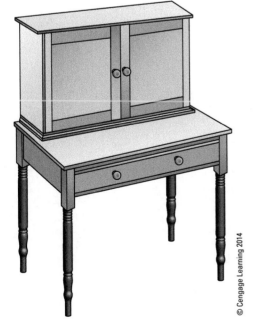

© Cengage Learning 2014

c. Top could be:

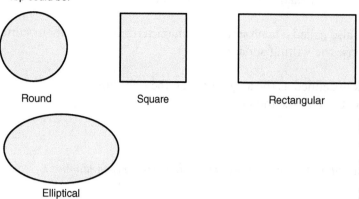

Round   Square   Rectangular

Elliptical

d.

e.

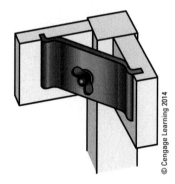

_____ 1. drop-leaf table

_____ 2. mitered tenons

_____ 3. postmaster's desk

_____ 4. tabletops

_____ 5. metal shopmade table brace

SHORT ANSWER

1. What are the steps involved in making and using a fixed tapering jig for a two-sided taper?

_____

_____

_____

_____

_____

_____

2. What standards are associated with dining tables?

_____

_____

_____

_____

_____

_____

3. How do you make button blocks?

_____

_____

_____

_____

_____

_____

_____

4. How is a compass used in making an eight-sided column?

_____

_____

_____

_____

_____

_____

_____

5. What configuration standards are associated with computer desks?

_____

_____

_____

_____

_____

_____

_____

_____

CHAPTER
23

Chests

MULTIPLE CHOICE
Identify the choice that best completes the statement or answers the question.

_____ 1. In its most primitive form, the blanket chest consists of just _____ boards.
   a. four
   b. five
   c. six
   d. seven

_____ 2. Which of the following is an example of a blanket chest?
   a. tool chest
   b. chest of drawers
   c. bureau
   d. dresser

_____ 3. The function of a chest is to _____.
   a. organize tools
   b. display items
   c. replace tables
   d. store items safely

_____ 4. Basswood is _____, and for this reason, it has been used extensively for kitchen utensils and food containers.
   a. citrus-scented
   b. heavy in weight
   c. odorless
   d. colorless

_____ 5. The chest of drawers is _____ than the blanket chest.
   a. older historically
   b. more difficult to build
   c. less complex
   d. less organizationally useful

_____ 6. The width of a chest-on-chest ranges from _____ inches.
   a. 18 to 24
   b. 36 to 48
   c. 72 to 84
   d. 96 to 108

_____ 7. A dresser is essentially a low and wide _____.
   a. chest-on-chest
   b. blanket chest
   c. chest of drawers
   d. armoire

_____ 8. The overall height of a dresser ranges from _____ inches.
   a. 29 to 34
   b. 39 to 44
   c. 49 to 54
   d. 59 to 64

_____ 9. Another name for an armoire is a _____, and it is a large, often ornate piece of furniture.
   a. blanket chest
   b. chest of drawers
   c. bureau
   d. wardrobe

_____ 10. When making an ogee bracket foot, the first step is to _____.
   a. sand the lumber
   b. make a template of the foot
   c. mill an appropriately sized strip of lumber
   d. draw the ogee on the piece of lumber

COMPLETION

Complete each statement.

1. The _____ chest is the simplest form of chest.

2. The dimensions of blanket chests vary greatly, but they typically run from 30 to _____ inches in length.

3. The _____ represents a much more evolved kind of chest; historically, it is a later form. As its name indicates, it is essentially a chest holding a number of drawers.

4. _____, _Tilia americana_, is also known as _American linden_ or _American lime_.

5. When making a(n) _____ scrolled cut, you only need to make a pattern for half of it.

6. A(n) _____ is just what it sounds like: One chest is placed on top of another.

138

7. When _____ on the table saw, resist the impulse to reach over with your hand to remove the cutoff; wait until the blade has stopped spinning.

8. A dresser is sometimes called a(n) _____.

9. It is hard to imagine, but houses were once built without closets. People used a piece of furniture called a(n) _____ to hold their hanging clothes.

10. The _____ into which the wedge is driven locks the mallet's handle firmly in the mallet head.

IDENTIFICATION

Identify each item pictured below. Write the letter of the best answer on the line next to each number.

a.

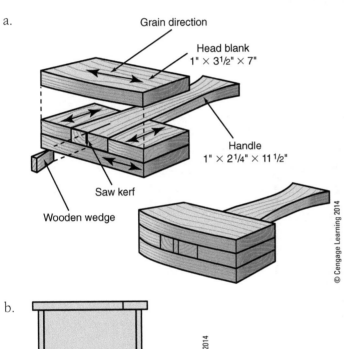

Grain direction

Head blank
1" × 3½" × 7"

Handle
1" × 2¼" × 11½"

Saw kerf

Wooden wedge

© Cengage Learning 2014

b.

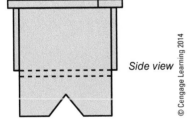

Side view

© Cengage Learning 2014

c.

d.

e.

15"–24"

12"–24"

© Cengage Learning 2014

_____ 1. blanket chest guidelines

_____ 2. primitive chest side view

_____ 3. block-front chest

_____ 4. wooden mallet diagram

_____ 5. dovetailed chest with tray

SHORT ANSWER

1. What are the general dimensions of a blanket chest?

_____

_____

_____

_____

_____

_____

2. What are the steps in making an applied base?

_____

_____

_____

_____

_____

3. What guidelines are followed to make a functional and attractive chest of drawers?

_____

_____

_____

_____

_____

4. How do you make a scrolled base?

_____

_____

_____

_____

_____

5. When making an ogee bracket foot, what steps are left after you rip the blank to final width?

_____

_____

_____

_____

_____

142

CHAPTER

24

Beds

MULTIPLE CHOICE

Identify the choice that best completes the statement or answers the question.

_____ 1. A bedstead is also called a _____.
- a. bedframe
- b. head rail
- c. headboard
- d. footboard

_____ 2. Bedposts may be square, turned, or _____.
- a. rectangular
- b. elliptical
- c. relieved
- d. tapered

_____ 3. The _____ rails run the length of the bed between the posts.
- a. head
- b. foot
- c. end
- d. side

_____ 4. How many posts does a bed have?
- a. one
- b. two
- c. three
- d. four

_____ 5. When the headboard and footboard are permanently connected to other parts of the bed, the headboard and footboard assemblies are usually _____ together.
- a. screwed
- b. mortised
- c. glued
- d. nailed

143

6. Which of the following is a standard for bed construction?
   a. The bed should be easily transportable.
   b. The box springs and mattress must be accommodated.
   c. Storage area must be provided.
   d. The bed must be aesthetically pleasing.

7. The recommended height from the floor to the top of the mattress is ____ inches.
   a. 12 to 15
   b. 17 to 20
   c. 24 to 27
   d. 30 to 33

8. Waterbeds generally use a ____-style bedstead.
   a. posted
   b. Murphy
   c. platform
   d. futon

9. The platform of a platform bed is often made of ____.
   a. plywood
   b. hardwood
   c. pine
   d. straw

10. A bed with tall and thin posts is known as a ____ bed.
    a. pencil-post
    b. banister
    c. futon
    d. sleigh

COMPLETION

Complete each statement.

1. The framework of a bed is called a(n) _____.

2. The posts at the bottom of the bed are connected by the foot rail and _____, if there is one.

3. _____, *Juglans cinerea,* is a member of the walnut family and is sometimes called white *walnut.*

4. The classic bed form has _____ to support the box springs.

5. The _____ is attached at the head of the bedstead.

6. The two posts at the head of the bed are connected by a head _____ and the headboard.

7. The simplest form of bed is the _____ bed.

8. _____ beds are a combination of sofa and bed.

9. Beds that are stacked one over the other are known as _____ beds.

10. Beds that fold up vertically, so they are parallel to the wall when closed, are known as _____ beds.

IDENTIFICATION

Identify each item pictured below. Write the letter of the best answer on the line next to each number.

a.

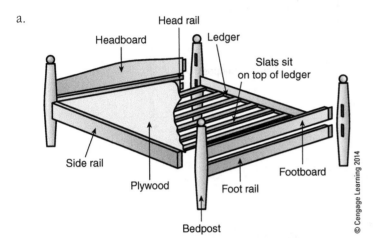

Head rail

Headboard

Ledger

Slats sit on top of ledger

Side rail

Plywood

Footboard

Foot rail

Bedpost

© Cengage Learning 2014

b.

© Cengage Learning 2014

c.

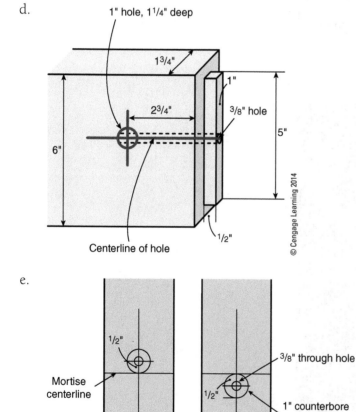

Side rail

3¼"

1¾"

6"

Centerline
of bolt

Nut

1" hole,
1¼" deep

Tenon-1" thick,
5" wide, ½" long

Inside view

Mortises
1" × 5", 9/16" deep

3/8" × 6"
bed bolt

9" or as indicated
on plans

2¾" square bedpost

© Cengage Learning 2014

d.

1" hole, 1¼" deep

1¾"

1"

2¾"

3/8" hole

6"

5"

Centerline of hole

½"

© Cengage Learning 2014

e.

½"

Mortise
centerline

Side 3

Vertical
centerline

3/8" through hole

½"

1" counterbore
¾" deep

Side 4

© Cengage Learning 2014

146

© 2014 Cengage Learning. All Rights Reserved. May not be scanned, copied or duplicated, or posted to a publicly accessible website, in whole or in part.

_____ 1. basic bed anatomy

_____ 2. carrying the centerline

_____ 3. centering a male bracket

_____ 4. drilling holes for bed bolt installation

_____ 5. layout for two bed bolts per post

SHORT ANSWER

1. Describe the characteristics of butternut wood.

_____

_____

_____

_____

_____

_____

_____

_____

_____

2. What are the first few steps involved in installing bed bolts (until you have to drill counterbore holes)?

_____

_____

_____

_____

_____

_____

_____

_____

_____

3. After drilling counterbore holes, what do you need to do to install bed bolts?

_____

_____

_____

_____

_____

_____

4. How do you install mortised bed rail fasteners?

_____

_____

_____

_____

_____

_____

5. How do you make a zero-clearance throat plate? Why is this important?

_____

_____

_____

_____

_____

_____

_____

CHAPTER

25

Chairs

MULTIPLE CHOICE

Identify the choice that best completes the statement or answers the question.

_____ 1. Chairs are divided into _____ main categories.
   a. two
   b. three
   c. four
   d. five

_____ 2. A side chair should have a seat height of _____ inches.
   a. 9 to 14
   b. 14 to 19
   c. 19 to 24
   d. 24 to 29

_____ 3. The curved pieces of wood attached to a rocking chair are known as _____.
   a. rockers
   b. rollers
   c. slats
   d. easers

_____ 4. A chair should _____ for optimum comfort.
   a. slant slightly forward
   b. slant slightly back
   c. be wider at the back
   d. not have arms

_____ 5. Chairs have two major components: the _____ and the supporting frame.
   a. spindles
   b. seat
   c. rungs
   d. slats

_____ 6. _____ extend from one rung to another, providing one means of tying the chair frame together.
   a. Crests
   b. Spindles
   c. Stretchers
   d. Splats

_____ 7. A seat to which the legs are attached is called a(n) _____ seat.
   a. rung
   b. frame
   c. open
   d. slab

_____ 8. As with chairs, bench parts may be rectangular, turned, or _____.
   a. mortised
   b. round
   c. square
   d. elliptical

_____ 9. Small couches are known as _____.
   a. loveseats
   b. folders
   c. pews
   d. stools

_____ 10. The first _____ was designed by Thomas Lee in 1903.
   a. porch swing
   b. sofa
   c. pew
   d. Adirondack chair

COMPLETION

Complete each statement.

1. A(n) _____ is defined as a piece of furniture, consisting of a seat, legs, back, and sometimes arms, that was designed to accommodate one person.

2. _____ chairs differ from armchairs in that they do not have arms.

3. _____ chairs are lower, both in seat height and overall height, than either side chairs or armchairs.

4. A(n) _____ chair is a chair with two curved pieces of wood attached to the bottom of the legs.

5. The _____ of a chair run between the legs and connect them.

6. The top of the back of a chair is called a(n) _____ rail.

7. _____ serve the same function as slats, but they are turned.

8. Both benches and _____ represent the earliest and simplest forms of seating that are still used today.

9. _____, also called *couches*, are actually larger versions of easy chairs.

10. _____ are essentially benches with backs, and like benches, they are built to seat more than a single occupant.

IDENTIFICATION

Identify each item pictured below. Write the letter of the best answer on the line next to each number.

a.

b.

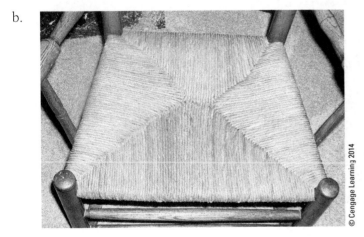

c.

d.

e.

© Cengage Learning 2014

____ 1. woven rush seat

____ 2. bench

____ 3. stool

____ 4. slab seat

____ 5. caned seat

SHORT ANSWER

1. Give a brief history of the chair.

_____

_____

_____

_____

_____

_____

_____

_____

_____

2. What are some characteristics of an easy chair?

_____

_____

_____

_____

_____

3. What types of joints are used in chairs?

_____

_____

_____

_____

_____

4. How can you more accurately drill holes in round stock?

_____

_____

_____

_____

_____

5. What are the characteristics of white ash?

_____

_____

_____

_____

_____

CHAPTER
26

Hardware

MULTIPLE CHOICE
Identify the choice that best completes the statement or answers the question.

_____ 1. An inset door is also known as a(n) _____ door.
a. flush
b. overlay
c. lipped
d. paneled

_____ 2. On butt hinges, _____ leaves are shaped for a closer fit.
a. flush
b. swaged
c. straight
d. inset

_____ 3. A _____ is an ornamental terminating point seen on a post, a piece of furniture, or a hinge.
a. straightedge
b. swage
c. pivot
d. finial

_____ 4. _____ hinges are similar to pivot hinges but are smaller and not as strong.
a. Knife
b. Flush
c. Finial
d. Formed

_____ 5. One type of invisible hinge is the _____ hinge; it is larger than the barrel hinge and has leaves that are mortised into the door edge and the cabinet.
a. butt
b. pin
c. pivot
d. Soss

155

_____ 6. European hinges are sometimes referred to as _____ hinges.
   a. cup
   b. Soss
   c. pin
   d. pivot

_____ 7. Drop-leaf _____ are required when using drop-leaf hinges.
   a. pins
   b. supports
   c. finials
   d. panels

_____ 8. _____ hold doors closed. They are located on the inside of the door and are not seen when the door is closed.
   a. Latches
   b. Hinges
   c. Handles
   d. Catches

_____ 9. A _____ is a vertically installed strip with slots to accept shelf supports, which in turn support the shelf.
   a. panel
   b. pilaster
   c. leaf
   d. finial

_____ 10. Related to levelers are _____, which may nail or screw on to the bottom of furniture, often the legs, to make it easy to move the furniture around without damaging the floor.
   a. extension slides
   b. lazy Susans
   c. escutcheons
   d. glides

COMPLETION

Complete each statement.

1. A(n) _____ is a jointed or flexible device that allows the turning or pivoting of a part on a stationary frame.

2. _____ hinges are also called _bent hinges_ or _wraparound hinges_.

3. _____ hinges consist of two plates that are riveted together. They can be used on overlay doors.

4. _____ hinges are completely hidden when the door is closed.

5. _____ hinges are concealed hinges that can be used with overlay and flush doors.

6. _____ and pulls are used for opening doors and drawers.

7. _____ are used on chests and trunks, and they enable a person to lift the trunk easily.

8. Shelf hardware consists of pilasters and shelf _____.

9. _____ allow a cabinet or table to be adjusted until it is level.

10. When table leaves are inserted into extension tables, they may be aligned with table _____ or aligned and secured with table locks.

IDENTIFICATION
Identify each item pictured below. Write the letter of the best answer on the line next to each number.

a.

© Cengage Learning 2014

b.

© Cengage Learning 2014

c.

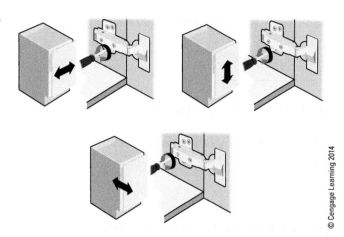

d.

e.

Section view

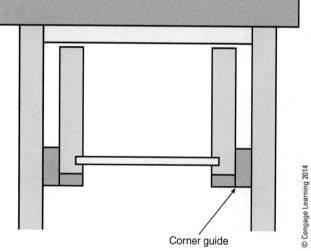

Corner guide

_____ 1. adjusting a European hinge

_____ 2. using a self-centering bit to drill pilot holes

_____ 3. pivot hinge

_____ 4. using a plywood spacer to set slides in cabinet

_____ 5. shop-built bottom slide

SHORT ANSWER

1. How do you install a butt hinge?

_____

_____

_____

_____

_____

_____

_____

_____

_____

_____

2. How do you install a formed hinge?

_____

_____

_____

_____

_____

_____

_____

_____

_____

_____

3. What is one way to simplify the installation of a piano hinge?

_____

_____

_____

_____

_____

_____

_____

4. How do you make multiple simple wooden pulls?

_____

_____

_____

_____

_____

_____

_____

5. How do you install recessed pilasters?

_____

_____

_____

_____

_____

_____

_____

CHAPTER
27

Surface Preparation

MULTIPLE CHOICE
Identify the choice that best completes the statement or answers the question.

_____ 1. _____ is soft limestone ground into a powder.
   a. A grinding wheel
   b. Rottenstone
   c. Pumice
   d. Aluminum oxide

_____ 2. _____ is rated Mohs 13; it is the hardest and most expensive of the synthetic abrasives.
   a. Silicon carbide
   b. Garnet
   c. Zirconia alumina
   d. Aluminum oxide

_____ 3. The first adhesive coat that binds abrasive grains to a backing is known as the _____ coat.
   a. flex
   b. grit
   c. bond
   d. size

_____ 4. One type of adhesive used to bind abrasive grains to a backing is called _____ over resin.
   a. resin
   b. glue
   c. animal hide
   d. waterproof

_____ 5. The grit size refers to the number of holes in the screen per lineal _____.
   a. centimeter
   b. inch
   c. foot
   d. meter

_____ 6. The lower the grit number, the _____ the abrasive grams are.
   a. finer
   b. harder
   c. softer
   d. coarser

_____ 7. For general sanding and easy leveling, _____ grit is a good choice of coated abrasive.
   a. 80
   b. 100
   c. 120
   d. 150

_____ 8. There are two kinds of scrapers: _____ scrapers and card scrapers.
   a. plane
   b. coarse
   c. sand
   d. cabinet

_____ 9. Deep scratches up to _____-inch deep can be removed with a hand plane.
   a. 1/16
   b. 1/8
   c. 1/4
   d. 1/2

_____ 10. _____-based putty is the most widely used.
   a. Lacquer
   b. Wax
   c. Oil
   d. Water

COMPLETION

Complete each statement.

1. The material that we call sandpaper is more accurately termed a coated _____.

2. _____ is finely ground lava.

3. Abrasive grains are rated on _____ scale, which rates minerals for hardness.

4. The second adhesive coat is called the _____ coat.

5. The _____ system is used to indicate the coarseness of the abrasive grains and consequently the coarseness of the sandpaper.

6. Dried _____ that is not removed will interfere with the wood's absorption of stain and finish.

7. Final sanding should always be done with the _____ of the wood.

162

8. _____ shears off wood fibers and creates a very smooth surface.

9. _____ involves removing wood around the defect and replacing it with another piece of wood.

10. Putties that harden may be either _____ or lacquer based.

IDENTIFICATION
Identify each item pictured below. Write the letter of the best answer on the line next to each number.

a.

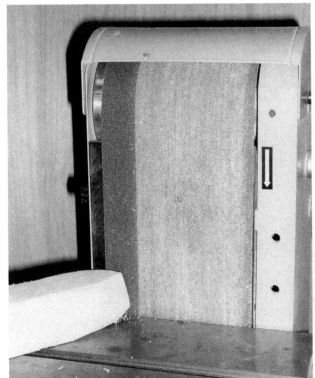

b.

c.

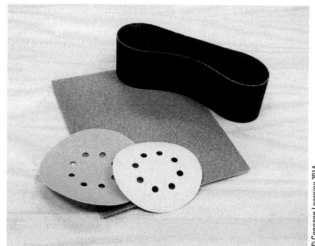

d.

e.

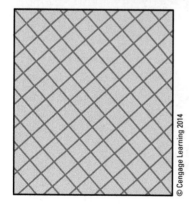

_____ 1. coated abrasives

_____ 2. cleaning a sanding belt

_____ 3. patching with a plug

_____ 4. double flex pattern on sandpaper

_____ 5. removing waste with a chisel

SHORT ANSWER

1. What types of backings are used with abrasive grains?

_____

_____

_____

_____

_____

_____

_____

_____

2. What is the purpose of flexing sandpaper?

_____

_____

_____

_____

_____

_____

_____

3. How do you sharpen a card scraper?

_____

_____

_____

_____

_____

_____

4. How do you patch a defect?

_____

_____

_____

_____

_____

_____

_____

_____

_____

5. Describe a shop-made alternative to wood putty.

_____

_____

_____

_____

_____

_____

_____

_____

_____

CHAPTER
28

Finishes and Application Methods

MULTIPLE CHOICE
Identify the choice that best completes the statement or answers the question.

____ 1. Applying finish to a project makes it ____.
   a. easier to stain
   b. easier to clean
   c. more liable to gather dust
   d. less beautiful

____ 2. The most commonly used water stains are ____ dyes.
   a. aniline
   b. oil
   c. turpentine
   e. spirit

____ 3. A strong ____ solution, 26 percent or higher, produces a soft brown stain.
   a. ammonia
   b. iron
   c. vinegar
   d. lye

____ 4. The great advantage of a ____ stain is that, due to the high viscosity, it will not run or drip.
   a. water
   b. chemical
   c. gel
   d. lacquer

____ 5. ____ is the strongest acting of the homemade bleaching solutions.
   a. Oxalic acid
   b. Chlorine laundry bleach
   c. Iron oxide
   d. Hydrogen peroxide

167

_____ 6. The classic nontoxic oil is _____ oil, but better choices are salad bowl oil and walnut oil
   a. olive
   b. almond
   c. mineral
   d. bamboo

_____ 7. The darkest grade of shellac is _____ lac.
   a. seed
   b. button
   c. garnet
   d. white

_____ 8. _____ is a synthetic varnish that is made up of plastics and solvents.
   a. Tung
   b. Polyurethane
   c. Lacquer
   d. Resin

_____ 9. The amount of sheen in polyurethane is controlled by the manufacturer by adding _____ to the mixture.
   a. VOCs
   b. lacquer
   c. shellac
   d. silica

_____ 10. LVHP stands for _____.
   a. low volume, high pressure
   b. low varnish, high pressure
   c. low varnish, high polyurethane
   d. low volume, high price

COMPLETION

Complete each statement.

1. _____ oil stains are made of coal-tar dyes dissolved in a thinner, which acts as a vehicle.

2. _____ stains are somewhat similar to water stains in that they are an aniline powder dye that is mixed with a vehicle; the difference is that the vehicle that the dye is dissolved in is denatured or ethyl alcohol or acetone rather than water.

3. _____ stains are made of either a penetrating oil stain or a pigmented oil stain to which wax and a drier have been added.

4. _____ stains are much thicker than any of the other stains; they have the consistency of jelly.

5. _____ oil is extracted from the seeds of the flax plant.

6. Tung oil is available as pure tung oil or as _____ tung oil.

7. _____ finishes, as their name implies, lie on the surface of the wood rather than penetrating into it.

8. Varnishes add a yellow tint, called _____ .

9. The development of water-based finishes was prompted by regulations concerning _____ (VOCs).

10. There are two types of spraying systems: One is a(n) _____ system, and the other is an HVLP system.

IDENTIFICATION

Identify each item pictured below. Write the letter of the best answer on the line next to each number.

a.

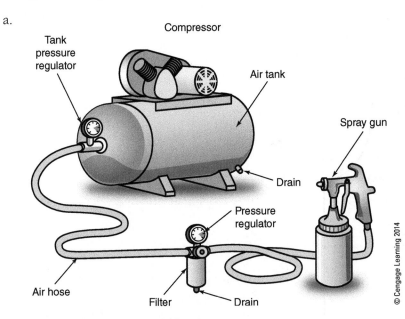

b.

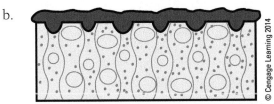

c.

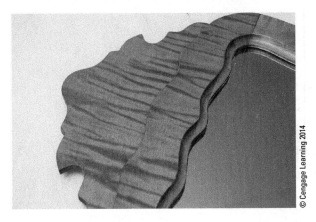

© 2014 Cengage Learning. All Rights Reserved. May not be scanned, copied or duplicated, or posted to a publicly accessible website, in whole or in part.

d.

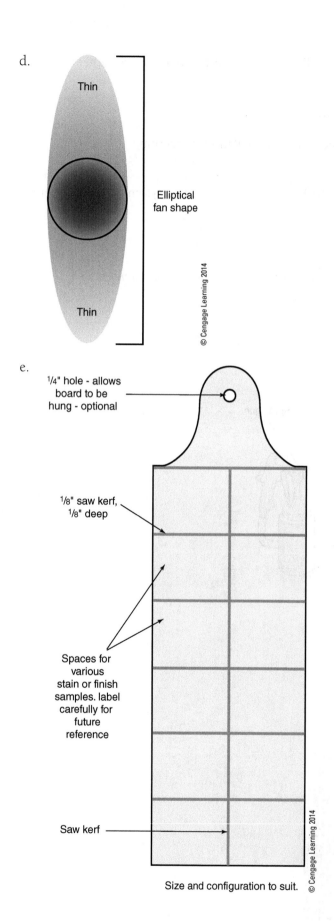

Thin

Elliptical
fan shape

Thin

e.

¼" hole - allows
board to be
hung - optional

⅛" saw kerf,
⅛" deep

Spaces for
various
stain or finish
samples. label
carefully for
future
reference

Saw kerf

Size and configuration to suit.

_____ 1. staining wood with a curly figure highlights grain

_____ 2. wood right after applying paste filler

_____ 3. stain-board diagram

_____ 4. compressor spraying system

_____ 5. adjusting a spray gun for a fan pattern

SHORT ANSWER

1. What sort of protection does finish provide?

_____

_____

_____

_____

_____

_____

_____

_____

2. What are the disadvantages of water stains?

_____

_____

_____

_____

_____

_____

_____

_____

_____

3. What are some characteristics of lacquer stains?

_____

_____

_____

_____

_____

4. Where does shellac come from, and how is it used?

_____

_____

_____

_____

_____

_____

5. What are combination finishes?

_____

_____

_____

_____

CHAPTER
29

Bending Wood

MULTIPLE CHOICE
Identify the choice that best completes the statement or answers the question.

_____ 1. Plain bending is limited to _____ stock.
   a. fairly thin pieces of
   b. thick pieces of
   c. hardwood
   d. pinewood

_____ 2. The _____ being bent will determine the thickness of the pieces you cut for laminations.
   a. price of the wood
   b. total length of the curve
   c. radius of the curve
   d. origin of the wood

_____ 3. The glues used to glue up bent laminations should be _____ glues.
   a. natural
   b. synthetic
   c. slow-set
   d. quick-set

_____ 4. Steam is invisible and highly dangerous. The steam generated for steam bending is at _____ °F.
   a. 0
   b. 32
   c. 100
   d. 212

_____ 5. Two-sided forms cannot be made by simply cutting the form material in two. Two _____ are required to allow for the thickness of the wood being bent.
   a. rectangles
   b. parallel lines
   c. perpendicular lines
   d. ellipses

_____ 6. Drying forms, called _____, are placed in the bent wood so that it keeps its shape as it dries.
   a. solid forms
   b. lamina
   c. dryers
   d. keepers

_____ 7. Regardless of the wood chosen for bending, it is important that it be free of defects and that it have a(n) _____ gram.
   a. straight
   b. end
   c. arced
   d. oblique

_____ 8. Segment lamination is also called _____.
   a. brick stacking
   b. milling
   c. kerfing
   d. coopering

_____ 9. In segment lamination, pieces are joined _____.
   a. edge to edge
   b. end to end
   c. front to front
   d. front to end

_____ 10. Coopering is an ancient trade. Coopers were responsible for making _____.
   a. shoes
   b. boats
   c. saddles
   d. casks

COMPLETION

Complete each statement.

1. _____ is a second method of dry bending that allows greater bends to be achieved than those possible through plain bending, and thicker pieces of wood may be used.

2. _____ bends are made by bending thin layers of stock that have been coated with glue around a form.

3. In lamination bending, the layers, called _____, are visible in the finished product.

4. The lignin in wood is made soft and pliable through _____, which allows the cellulose fibers bound together by the lignin to slide past one another.

5. _____ bending is sometimes called _hot-pipe bending_ and is most frequently associated with the construction of musical instruments.

6. It is also possible to plasticize wood and then bend it with very hot or boiling _____.

7. The United States _____ Laboratory has conducted bending tests on hundreds of boards from many different species.

8. _____ lamination is a method for making curves using short pieces of solid stock stacked in staggered rows.

9. _____ is one method of creating curves without bending; it is a method of joining wood strips edge to edge to create a curve.

10. The beveled pieces used in coopering are called _____.

IDENTIFICATION

Identify each item pictured below. Write the letter of the best answer on the line next to each number.

a.

b.

c.

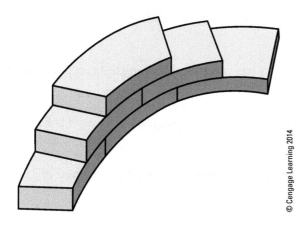

d.

Top view

A piece of plywood is used for the bottom

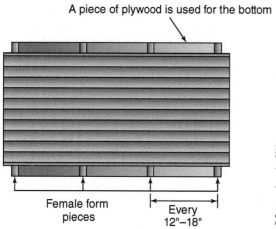

Female form
pieces

Every
12"–18"

e.

____ 1. curved chair back

____ 2. dry-heat bending jig

____ 3. keepers

____ 4. gluing form for coopered panels

____ 5. segment lamination

SHORT ANSWER

1. How do you determine kerf spacing for a given bend?

_____

_____

_____

_____

_____

_____

_____

2. Can you perform dry-heat bending with a hot air gun?

_____

_____

_____

_____

_____

_____

_____

3. When building a steam box, what steps should be followed before gluing the plug to the door?

_____

_____

_____

_____

_____

_____

_____

4. How can you mill curves from solid lumber? Is it cost-effective?

_____

_____

_____

_____

_____

_____

_____

_____

5. When creating a curve using segment lamination, what steps should you follow before you assemble the first row of pieces?

_____

_____

_____

_____

_____

_____

_____

_____

_____

178

CHAPTER

30

Veneering

MULTIPLE CHOICE

Identify the choice that best completes the statement or answers the question.

_____ 1. Veneer is cut with a veneer saw, which is designed to be used against a _____ in the same way that a knife is.
- a. veneer punch
- b. piece of veneer tape
- c. straightedge
- d. veneer hammer

_____ 2. The surface to which the veneer is attached is called the *substrate,* or *core material;* it is also referred to as the _____.
- a. edge band
- b. groundwork
- c. punch
- d. film

_____ 3. Arranging leaves that came from the same flitch side by side is called a _____ pattern.
- a. slip
- b. book
- c. diamond
- d. cross

_____ 4. If leaves from the same flitch are positioned like they were opened from a book, the resulting pattern is called a book _____.
- a. match
- b. band
- c. cross
- d. slip

_____ 5. Cross banding generally has a grain that runs at a _____ angle to the edge of the central veneer panel.
- a. 30°
- b. 45°
- c. 60°
- d. right

_____ 6. When the veneer is sliced from the log, tiny checks are created on one face; this is the _____ face.
   a. tight
   b. cross
   c. book
   d. open

_____ 7. When cauls are used, _____ adhesives are employed.
   a. natural
   b. synthetic
   c. slow-setting
   d. quick-setting

_____ 8. One option for adhesive is _____ cement, which is brushed or rolled onto the core material and onto the back of the veneer and allowed to dry.
   a. contact
   b. portland
   c. film
   d. hide

_____ 9. To repair a blister or lump in veneer, you should cut a(n) _____ -shaped flap with a sharp knife.
   a. V
   b. X
   c. circle
   d. square

_____ 10. Veneer edge trimmers have _____ sides.
   a. one or two
   b. three or four
   c. five
   d. six

COMPLETION

Complete each statement.

1. Veneer _____ is used to hold the cut veneer pieces together until after they are glued.

2. A(n) _____ hammer is not a hammer in the traditional sense; it is used to press the veneer down against the core material once the adhesive is in place.

3. To encourage good adhesion, solid wood substrates are roughed with a(n) _____ plane, which has grooves in the plane iron.

4. Veneer _____ are used to repair defects in veneer.

5. _____ matching and book matching are the two most common ways to match veneer, but many other possibilities exist.

6. Cross _____ is a technique used to set off a center panel; it is essentially a border, and it has the effect of highlighting the center section.

7. Veneer has a front and a back, more properly referred to as a(n) _____ face and an open face.

8. Attaching veneer to groundwork is done either by hand or with _____, sturdy boards between which the glued veneer and substrate are clamped.

9. A(n) _____ is a flat panel that has been kerfed to a depth of about 1/8" at 2" intervals along both its length and width.

10. The oldest method of laying veneer employs _____ glue.

IDENTIFICATION
Identify each item pictured below. Write the letter of the best answer on the line next to each number.

a.

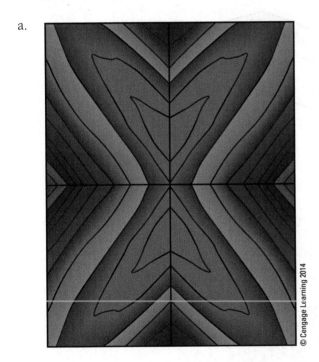

© Cengage Learning 2014

b.

c.

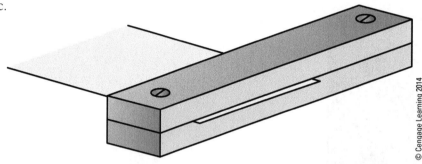

d.

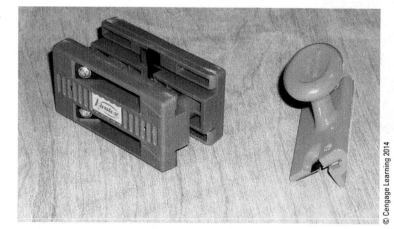

e.

© Cengage Learning 2014

_____ 1. using a veneer saw

_____ 2. toothing plane

_____ 3. reverse-diamond match

_____ 4. clamping jig for trimming veneer

_____ 5. veneer edge trimmers

SHORT ANSWER

1. What is the history of veneering?

_____

_____

_____

_____

_____

_____

_____

_____

2. How should you handle and store veneer?

_____

_____

_____

_____

_____

3. How do you flatten veneer?

_____

_____

_____

_____

_____

_____

_____

_____

4. How is glue film used in attaching veneer?

_____

_____

_____

_____

_____

_____

_____

_____

5. What are the characteristics of veneer used in edge banding?

_____

_____

_____

_____

_____

_____

_____

CHAPTER
31

Decorative Techniques

MULTIPLE CHOICE
Identify the choice that best completes the statement or answers the question.

_____ 1. Inlay involves decorating a surface _____ to the work.
   a. by adding thickness
   b. by adding paint
   c. without making alterations
   d. without adding any thickness

_____ 2. _____ are made up of two or more species of wood; they have a design, which may be very intricate.
   a. Bandings
   b. Strings
   c. Scratch stocks
   d. Router planes

_____ 3. If the same procedure used to create the checkerboard pattern is followed, but the grain of every other strip is reversed, the end result is a _____ design.
   a. marquetry
   b. basket-weave
   c. chessboard
   d. lettered

_____ 4. Diamond patterns can be created by cutting an alternating pattern at a _____-degree angle.
   a. 30
   b. 60
   c. 90
   d. 135

_____ 5. A classic pattern is the _____, created by joining three diamonds.
   a. checkerboard
   b. basket-weave
   c. isometric cube
   d. rhombus

185

_____ 6. Whether a scroll saw or a fret saw is used to cut marquetry patterns, a _____ blade is required.
   a. thick, fine-toothed
   b. thick, coarse-toothed
   c. thin, fine-toothed
   d. thin, coarse-toothed

_____ 7. The best solution for finishing inlay is to use a very fine grit abrasive, _____ or higher, and to use a random oscillating sander.
   a. 80
   b. 120
   c. 180
   d. 220

_____ 8. In _____, the background is cut away to leave a design.
   a. chip carving
   b. relief carving
   c. carving in the round
   d. geometric carving

_____ 9. The _____ is the primary carving tool in chip carving.
   a. fret saw
   b. band saw
   c. cutting knife
   d. stab knife

_____ 10. Letters you wish to carve should be transferred to the cutting surface with _____.
   a. carbon paper
   b. tracing paper
   c. a photocopier
   d. a stab knife

COMPLETION

Complete each statement.

1. A(n) _____ is a tool that holds a sharp cutter.

2. _____ are thin, narrow strips of wood from one species.

3. _____ differs from inlay work in that the material set into the recess created in the ground surface is not flush with the surface but stands proud of it.

4. _____ is the creation of geometric patterns or motifs through the use of symmetrically shaped pieces of veneer.

5. A(n) _____ saw is like a coping saw but has a much deeper throat.

6. One method of cutting marquetry designs is called the overlapping method, or _____ method.

7. Cutting pieces of veneer individually for a picture can be done with a sharp knife, using a technique called the _____ method.

8. Carvings fall into one of three categories: relief carving, carving in the _____, and chip carving.

9. There are two basic knives used in chip carving, a cutting knife and a(n) _____ knife.

10. _____ letters are very legible and are perhaps the easiest type of lettering to cut.

IDENTIFICATION

Identify each item pictured below. Write the letter of the best answer on the line next to each number.

a.

b.

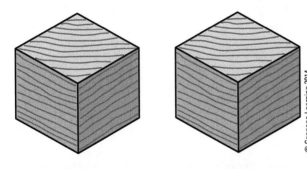

c.

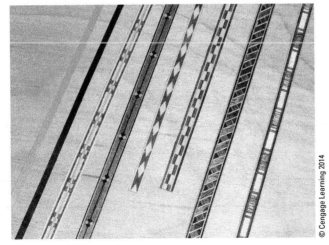

187

d.

e.

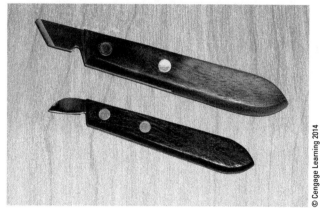

_____ 1. router plane and scratch stock

_____ 2. strings and bandings

_____ 3. isometric cube pattern

_____ 4. stab knife and cutting knife

_____ 5. rosettes

SHORT ANSWER

1. How do you make a scratch stock?

_____

_____

_____

_____

_____

_____

_____

2. Describe the checkerboard parquetry pattern.

_____

_____

_____

_____

_____

_____

3. How do you make a parquetry jig?

_____

_____

_____

_____

_____

_____

_____

_____

_____

_____

_____

_____

4. What is the purpose of shading veneer? Briefly describe the process of shading.

_____

_____

_____

_____

_____

_____

_____

_____

5. Which species of wood are most suitable for chip carving?

_____

_____

_____

_____

_____

_____

_____

_____

_____

_____

CHAPTER
32

Designing, Drawing, and Planning

MULTIPLE CHOICE
Identify the choice that best completes the statement or answers the question.

_____ 1. _____ may be straight, curved, circular, or S-shaped.
   a. Tones
   b. Masses
   c. Lines
   d. Shapes

_____ 2. Furniture or cabinetry exhibiting formal balance is _____.
   a. colorful
   b. symmetrical
   c. asymmetrical
   d. ornate

_____ 3. According to the rule of _____ progression, each successive unit within the frame increases by a constant ratio.
   a. geometric
   b. arithmetic
   c. Fibonacci
   d. harmonic

_____ 4. In a _____ mass, an object is taller than it is wide.
   a. secondary horizontal
   b. secondary vertical
   c. primary horizontal
   d. primary vertical

_____ 5. The Pilgrim style is sometimes called _____.
   a. Jacobean
   b. Elizabethan
   c. Tudor
   d. Victorian

_____ 6. The ____ style was popular from 1725 to 1755. Graceful and fluid curved lines characterized this style.
a. Baroque
b. Queen Anne
c. Postmodern
d. William and Mary

_____ 7. The ____ style, prevalent from 1690 to 1850, is a catch-all term for furniture produced outside the urban centers.
a. Country
b. Rural
c. Chippendale
d. Pennsylvania Dutch

_____ 8. The ____ style grew out of the International style following World War II; it features furniture that is versatile, economical, and lacking in ornamentation.
a. Postmodern
b. Baroque
c. Chippendale
d. Contemporary

_____ 9. Often included in a full set of drawings will be one or more _____ that show what the cabinet would look like if part of it were to be cut away.
a. cut lists
b. section views
c. orthographic projections
d. isometric drawings

_____ 10. A(n) ____ details how the piece is built in a step-by-step manner.
a. cut list
b. section view
c. plan of procedure
d. orthographic projection

COMPLETION

Complete each statement.

1. _____ is what the piece is designed to do, or its purpose.

2. _____ is the relationship between line and shape that gives the appearance of substance.

3. _____ may be defined as color quality, such as brightness, deepness, or hue.

4. _____ is the relationship of the parts of an object to each other and to the whole; examples are height and width relationships.

5. An example of a(n) _____ series is the series 1, 1, 2, 3, 5, 8, 13.

192

6. _____ take into account physical differences between people and objects.

7. The _____ period, which was prominent from 1780 to 1820, has been called the United States' first homegrown style.

8. Some designers, who wished to incorporate older Baroque and neoclassical elements for decoration, developed what is known as the _____ style.

9. _____ drawings are two dimensional; because they do not show depth, they do not appear "true" to our eye.

10. A(n) _____ list is a list of all the parts needed for a given project, the number of pieces needed, and the dimensions of each piece.

IDENTIFICATION

Identify each item pictured below. Write the letter of the best answer on the line next to each number.

a.

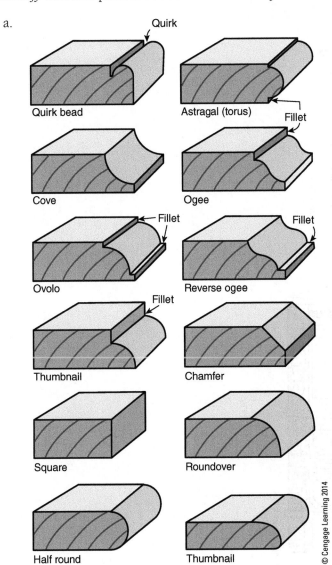

Quirk

Quirk bead

Astragal (torus)

Fillet

Cove

Ogee

Fillet

Ovolo

Fillet

Reverse ogee

Fillet

Thumbnail

Chamfer

Square

Roundover

Half round

Thumbnail

© Cengage Learning 2014

b.

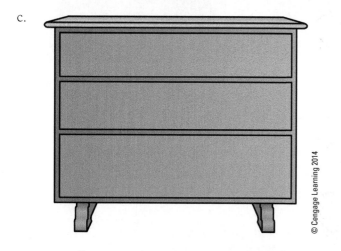

c.

d.

e.

_____ 1. edge profiles

_____ 2. informal balance

_____ 3. golden rectangle

_____ 4. sled feet

_____ 5. Pilgrim chest

SHORT ANSWER

1. What is the difference between form and function?

_____

_____

_____

_____

_____

_____

_____

_____

2. What is the difference between harmony and repetition?

_____

_____

_____

_____

_____

_____

_____

_____

3. What are standards?

_____

_____

_____

_____

_____

_____

_____

4. What were the hallmarks of Shaker design?

_____

_____

_____

_____

_____

_____

_____

_____

5. What are materials and supplies, and when do you need to obtain them?

_____

_____

_____

_____

_____

_____

_____

_____

_____

CHAPTER

33

Fabricating Countertops

MULTIPLE CHOICE

Identify the choice that best completes the statement or answers the question.

_____ 1. Granite is _____.
   a. flimsy looking
   b. expensive
   c. not very aesthetically pleasing
   d. unstained by cooking oil

_____ 2. _____ has long been used as a countertop material in science labs.
   a. Soapstone
   b. Granite
   c. Ceramic
   d. Marble

_____ 3. Wood must be sealed periodically with _____, which is nontoxic.
   a. WD-40
   b. acetone
   c. ethyl alcohol
   d. mineral oil

_____ 4. _____ is (are) the most commonly used material for kitchen countertops.
   a. Laminates
   b. Granite
   c. Soapstone
   d. Metals

_____ 5. Since countertops are typically _____ inch(es) thick, the substrate must be built up around its edges.
   a. 1
   b. 1 1/4
   c. 1 1/2
   d. 1 3/4

_____ 6. When cutting laminate with a portable power saw, it should be cut with the decorative side down to avoid _____.
   a. scorching
   b. scratching
   c. chipping
   d. melting

_____ 7. When laminating doors or drawer fronts, the _____ is (are) laminated first.
   a. back
   b. sides
   c. front
   d. bottom

_____ 8. Once the laminate has been securely attached to the substrate, the waste protruding past the edge of the surface is trimmed off; this is typically done with a laminate trimmer, using a(n) _____-cut bit.
   a. piloted protruding
   b. piloted flush
   c. automatic protruding
   d. automatic flush

_____ 9. One method of placing a drop edge on a solid surface countertop is known as a _____.
   a. single drop edge
   b. double drop edge
   c. vertical drop edge
   d. horizontal drop edge

_____ 10. Solid surface countertops can be _____ to the desired finish.
   a. painted
   b. sanded
   c. routed
   d. planed

COMPLETION

Complete each statement.

1. Countertops consist of a flat surface and often a(n) _____.

2. _____ tile is a product made from clay and other materials; tile is fired in a kiln to harden it.

3. _____ has become a very popular countertop material; a natural stone found around the world, it comes in a variety of colors and patterns.

4. _____ is a natural stone, and it is very expensive; it is waterproof, heatproof, and very beautiful.

5. _____ is a quarried natural stone; it is generally a rich, dark gray in color and very smooth to the touch.

6. _____ is a construction material that consists of Portland cement, gravel, sand, and water.

7. Hard _____ is the most commonly used wood for countertops due to its tight grain and exceptional hardness.

8. Plywood, particleboard, or MDF may be used as a(n) _____ for laminates.

9. When installing solid surface countertops, no _____ should be placed within an inch of any corner or within three inches of a dishwasher or stove.

10. A(n) _____ edge can be placed on the countertop using one of two methods. The stacking method is stronger and is preferred.

IDENTIFICATION

Identify each item pictured below. Write the letter of the best answer on the line next to each number.

a.

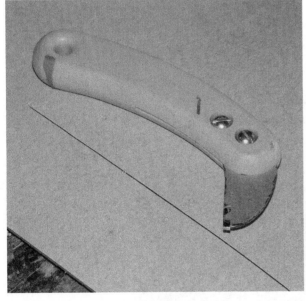

© Cengage Learning 2014

b.

c.

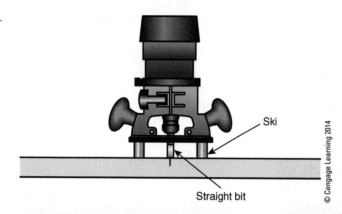

Ski

Straight bit

d.

e.

____ 1. granite countertop

____ 2. laminate scoring tool

____ 3. seam support

____ 4. chamfered edge

____ 5. removing squeeze-out

SHORT ANSWER

1. What are the characteristics of metal countertops?

_____

_____

_____

_____

_____

_____

_____

2. What are the disadvantages of solid surface countertops?

_____

_____

_____

_____

_____

_____

_____

3. What should you do if you need to apply laminate to a large area?

_____

_____

_____

_____

_____

_____

4. What reinforcement is necessary for solid surface countertops?

_____

_____

_____

_____

_____

_____

5. How do you attach solid surface countertops to cabinets? How do you install backsplashes?

_____

_____

_____

_____

_____

_____

_____

CHAPTER
34

Running Your Own Business

MULTIPLE CHOICE
Identify the choice that best completes the statement or answers the question.

_____ 1. No matter which business structure you choose, you will need to pay FICA taxes; these are the same as _____ taxes.
   a. Social Security
   b. federal income
   c. state income
   d. sales

_____ 2. Insurance considerations include _____ insurance, health insurance, and workers' compensation insurance, if you have employees.
   a. life
   b. auto
   c. tax
   d. liability

_____ 3. Most small businesses start out as _____.
   a. LLCs
   b. sole proprietorships
   c. corporations
   d. general partnerships

_____ 4. In a limited partnership, the _____ partner or partners do most of the decision making, but they also have full liability.
   a. marketing
   b. general
   c. managing
   d. founding

_____ 5. Although a corporation can be taxed or sued, its owners, called _____, are not personally liable for the corporation's debts.
   a. partners
   b. managers
   c. shareholders
   d. proprietors

_____ 6. A corporation is chartered by a _____.
   a. city
   b. county
   c. country
   d. state

_____ 7. In a business plan, the _____ section should include the reasons for starting and operating your business.
   a. production plan
   b. business objectives
   c. business form
   d. resume

_____ 8. In a business plan, the _____ section should include balance sheets.
   a. business references
   b. marketing plan
   c. equipment
   d. financial

_____ 9. In a business plan, the _____ section should include inventory.
   a. competition
   b. equipment
   c. marketing plan
   d. facilities

_____ 10. It is a good idea to have a _____ of your work to show potential customers.
   a. business card
   b. portfolio
   c. resume
   d. business plan

COMPLETION

Complete each statement.

1. A(n) _____ is a person who assumes the responsibility and the risk for a business operation with the expectation of making a profit.

2. _____ is the task of promoting and selling your product.

3. The simplest form of business structure is a(n) _____.

4. In a(n) _____ structure, two or more people own the business.

5. When profits are shared among the partners, it is known as a(n) _____ partnership.

6. Unlike a sole proprietorship or a partnership, a(n) _____ is considered an entity unto itself, apart from the people who own it.

7. The _____ company is a relatively new business structure that is a hybrid between a partnership and a corporation.

8. There are two main types of corporations, C corporations and _____ corporations.

9. As you first ponder the idea of starting a business, begin by jotting down ideas to answer the questions what, when, where, why, and _____.

10. Your _____ should clearly state what your business is, who you are, and how to contact you.

SHORT ANSWER

1. What skills do successful business owners share?

_____

_____

_____

_____

_____

_____

_____

_____

_____

2. What questions should you ask yourself when deciding whether or not to start a business?

_____

_____

_____

_____

_____

_____

_____

_____

_____

3. What are the advantages and disadvantages of a sole proprietorship?

_____

_____

_____

_____

_____

_____

_____

_____

4. What should you keep in mind when you sell at galleries or craft shows?

_____

_____

_____

_____

_____

_____

_____

_____

5. What is the importance of good record keeping?

_____

_____

_____

_____

_____

_____

_____

_____

PART II

SKILL SHEETS

PROCEDURE 3-1: Using a Handsaw

Objective: The student shall successfully demonstrate the use of a handsaw.
Materials Needed: Board
Square
Pencil
Handsaw
Safety Requirements: Eye protection
References: *Chapter 3, Hand Tools*
Procedure 3-1, p. 37

Student Name: _____ **Date:** _____

Evaluator Name: _____ **Date:** _____

Ratings: 0 = Skipped Task; 1 = Attempted, but did not complete Task; 2 = Poor demonstration of Task;
3 = Fair demonstration of Task; 4 = Good demonstration of Task; 5 = Outstanding demonstration of Task.

Task	0	1	2	3	4	5
Wear proper safety equipment.						
Using a square, mark a line on the finish side of your material.						
Mark the waste side of the line with an X.						
Start the cut by holding the handle of the saw with your index finger pointing down and the thumb of your other hand holding the blade in place.						
The cut is made on the waste side of the line.						
Start the cut with a downward motion of the saw, removing your thumb once the cut is established.						
Cut through the board with long steady strokes.						
TOTAL POINTS						
FINAL (AVERAGED) SCORE						

EVALUATOR NOTES	STUDENT NOTES

PROCEDURE 3-2: Using a Chisel to Cut a Hinge Recess

Objective: The student shall successfully use a chisel to cut a hinge recess.
Materials Needed: Board
Knife
Hinge
Chisels
Mallet
Safety Requirements: Eye protection
References: *Chapter 3, Hand Tools*
Procedure 3-2, p. 40

Student Name: _____ Date: _____

Evaluator Name: _____ Date: _____

Ratings: 0 = Skipped Task; 1 = Attempted, but did not complete Task; 2 = Poor demonstration of Task; 3 = Fair demonstration of Task; 4 = Good demonstration of Task; 5 = Outstanding demonstration of Task.

Task	0	1	2	3	4	5
Wear proper safety equipment.						
Secure the material that you will be working on with clamps or in a vise.						
Mark out the area to be chiseled using the hinge as a guide and scoring your layout lines with a knife.						
Mark the depth of the recess on the edge of the board.						
Position a chisel that is about the same width as the recess on the scored line with the bevel facing in to the recess; strike it with a mallet, cutting to the full depth of the recess.						
Use a wider chisel to do the same thing along the back of the recess.						
Take your first chisel and make a series of cuts about a quarter inch apart from one end of the outline to the other.						
Place a chisel on the edge of the board along the marked depth line, bevel facing up, and cut away the waste.						
Fit hinge in recess.						
TOTAL POINTS						
FINAL (AVERAGED) SCORE						

EVALUATOR NOTES	STUDENT NOTES

PROCEDURE 3-3: Setting Up and Using a Bench Plane

Objective: The student shall successfully demonstrate setting up and using a bench plane.
Materials Needed: Bench plane
Board
Safety Requirements: Eye protection
References: *Chapter 3, Hand Tools*
Procedure 3-3, p. 43

Student Name: _____ Date: _____

Evaluator Name: _____ Date: _____

Ratings: 0 = Skipped Task; 1 = Attempted, but did not complete Task; 2 = Poor demonstration of Task;
3 = Fair demonstration of Task; 4 = Good demonstration of Task; 5 = Outstanding demonstration of Task.

Task	0	1	2	3	4	5
Wear proper safety equipment.						
Release the cap lock and remove the lever cap from the plane.						
Take out the cap iron and the blade; they are held together with the cap iron screw.						
The blade should extend about 1/16" beyond the end of the cap iron. If this needs to be adjusted, remove the cap iron screw, reposition the cap iron, and replace the cap iron screw.						
Replace the blade assembly in the plane and secure it in place with the lever cap.						
Check the gap between the front of the plane iron and the front of the mouth. It should be between 1/32" and 1/16". If the gap is too wide or too narrow, remove the blade assembly.						
Loosen the two locking screws and then adjust the frog adjuster screw to set the proper gap and tighten the locking screws.						
Reposition the blade assembly on the frog, and lock it in place with the cap lever.						
Hold the plane upside down and check to see that the edge of the plane iron is level and centered in the mouth. If it is not, adjust it with the lateral adjustment lever.						
Adjust your plane for depth of cut by releasing the cap lock and advancing the blade with the depth of adjustment knob so that it protrudes from the mouth about 1/32" and reset the cap lock.						
Secure a scrap piece and make a test cut. Hold the plane with two hands; start the cut with pressure on the front of the plane, and end it with pressure on the back of the plane.						
Keep the sole of the plane flat against the work and plane with firm, constant pressure.						
Plane with the grain of the wood.						
TOTAL POINTS						
FINAL (AVERAGED) SCORE						

EVALUATOR NOTES	STUDENT NOTES

PROCEDURE 3-4: Sharpening Chisels and Plane Irons Using a Sandpaper System

Objective: The student shall successfully demonstrate sharpening chisels and plane irons using a sandpaper system.

Materials Needed: Chisel(s) and/or bench plane(s)
3/4" MDF or 1/4" plate glass for sandpaper substrate
Self-adhesive sandpaper in various grits or regular sandpaper and spray-mount adhesive
Honing jig (optional)

Safety Requirements: None

References: *Chapter 3, Hand Tools*
Procedure 3-4, p. 46

Student Name: _____ **Date:** _____

Evaluator Name: _____ **Date:** _____

Ratings: 0 = Skipped Task; 1 = Attempted, but did not complete Task; 2 = Poor demonstration of Task; 3 = Fair demonstration of Task; 4 = Good demonstration of Task; 5 = Outstanding demonstration of Task.

Task	0	1	2	3	4	5
Attach the sandpaper to the substrate and label each grit.						
Start with the backs of your blades, flattening and polishing the first inch or so by working through the series of grits, starting with the coarsest.						
Keeping the back flat on the abrasive surface, move it back and forth until it is uniformly scratched, and then move to the next grit.						
Begin to work on the bevel. Jigs are available to hold the blade at the desired angle.						
Work your way through the same progression of abrasives that you used on the back. Work each grit until you feel an even wire along the edge.						
Smooth the wire edge off by taking a couple of strokes with the back of the tool on the sharpening medium, and move to the next higher grit.						
When you are done, you should have a mirror finish on both the front and back of your blade.						
TOTAL POINTS						
FINAL (AVERAGED) SCORE						

EVALUATOR NOTES	STUDENT NOTES

PROCEDURE 4-1: Setting Cutting Depth on a Circular Saw

Objective: The student shall successfully demonstrate setting cutting depth on a circular saw.

Materials Needed: Circular saw
Board

Safety Requirements: None, as there is no cutting

References: *Chapter 4, Portable Power Tools*
Procedure 4-1, p. 67

Student Name: _____   **Date:** _____

Evaluator Name: _____   **Date:** _____

Ratings: 0 = Skipped Task; 1 = Attempted, but did not complete Task; 2 = Poor demonstration of Task; 3 = Fair demonstration of Task; 4 = Good demonstration of Task; 5 = Outstanding demonstration of Task.

Task	0	1	2	3	4	5
With the saw unplugged, retract the lower blade guard and set the base plate on the work piece, butting the blade against the edge of the stock.						
Determine whether your saw is a pivot-foot or a drop-foot model.						
If it is a pivot-foot saw, release the depth adjustment lever. Then, keeping the base flat on the material, hold the handle and pivot the saw up and down.						
If your saw is a drop-foot model, loosen the depth adjustment knob, then hold the base plate steady as you pull up or press down on the handle.						
Adjust the saw until the blade clears the stock by about 1/4". One tooth, and at least part of the adjoining gullets, should project below the material.						
Lock the lever on a pivot-foot saw or tighten the knob on a drop-foot saw.						
TOTAL POINTS						
FINAL (AVERAGED) SCORE						

EVALUATOR NOTES	STUDENT NOTES

PROCEDURE 4-2: Making Curved Cuts with a Jigsaw

Objective: The student shall successfully demonstrate making curved cuts using a jigsaw.
Materials Needed: Jigsaw
Board
Safety Requirements: Eye protection
References: *Chapter 4, Portable Power Tools*
Procedure 4-2, p. 68

Student Name: _____ **Date:** _____

Evaluator Name: _____ **Date:** _____

Ratings: 0 = Skipped Task; 1 = Attempted, but did not complete Task; 2 = Poor demonstration of Task; 3 = Fair demonstration of Task; 4 = Good demonstration of Task; 5 = Outstanding demonstration of Task.

Task	0	1	2	3	4	5
Wear proper safety equipment.						
Clamp the material to be cut to a support, letting the area to be cut to overhang the support.						
Make release cuts by cutting in from the edge of the material into the tightest turns.						
Starting at the end of the board, feed the saw into the material, guiding the saw to keep it on the line.						
Saw to the first release cut; the waste will fall away.						
Make successive cuts to each of the release cuts.						
Complete the job by sawing back from the opposite end of each line into each release cut.						
Finish cutting the curves that remain uncut.						
TOTAL POINTS						
FINAL (AVERAGED) SCORE						

EVALUATOR NOTES	STUDENT NOTES

PROCEDURE 4-3: Drilling Straight and Angled Holes Accurately

Objective: The student shall successfully demonstrate drilling straight and angled holes accurately.

Materials Needed: Drill bits
Try square
Sliding T-bevel
Board
Thick scrap of wood

Safety Requirements: Eye protection

References: *Chapter 4, Portable Power Tools*
Procedure 4-3, p. 74

Student Name: _____ Date: _____

Evaluator Name: _____ Date: _____

Ratings: 0 = Skipped Task; 1 = Attempted, but did not complete Task; 2 = Poor demonstration of Task; 3 = Fair demonstration of Task; 4 = Good demonstration of Task; 5 = Outstanding demonstration of Task.

Task	0	1	2	3	4	5
Wear proper safety equipment.						
Drill a straight hole accurately by keeping the drill bit perpendicular to the material being drilled. Make a guide block by cutting a 90° wedge out of one corner of a thick board.						
Center the bit in the drill over the mark to be drilled.						
Butt the notched corner of the guide block to the bit and clamp the block in place.						
Keep the bit flush against the corner of the block and bore the hole.						
Drill a straight hole using a try square as a guide by keeping the bit parallel to the square as you drill.						
To drill an angled hole, use a sliding T-bevel to guide the drill bit. Set the sliding T-bevel to the desired angle.						
Line up its handle beside the point where you need to drill the hole.						
Center the bit over the mark and bore the hole, keeping the bit parallel to the blade of the sliding T-bevel while you drill.						
TOTAL POINTS						
FINAL (AVERAGED) SCORE						

EVALUATOR NOTES	STUDENT NOTES

PROCEDURE 4-4: Shaping an Edge with a Piloted Router Bit

Objective:	The student shall successfully demonstrate shaping an edge with a piloted router bit.
Materials Needed:	Router and wrenches
	Piloting edging bit
	Board
Safety Requirements:	Eye protection
	Hearing protection
References:	*Chapter 4, Portable Power Tools*
	Procedure 4-4, p. 79

Student Name: _____   Date: _____

Evaluator Name: _____   Date: _____

Ratings: 0 = Skipped Task; 1 = Attempted, but did not complete Task; 2 = Poor demonstration of Task; 3 = Fair demonstration of Task; 4 = Good demonstration of Task; 5 = Outstanding demonstration of Task.

Task	0	1	2	3	4	5
Wear proper safety equipment.						
Select the proper piloted bit for the edge design desired.						
Secure the bit in the router's collet, and use the wrenches to tighten the collet nut.						
Adjust the depth of the cut. Deep cuts should be made in successive shallow passes.						
Secure the material to a bench with clamps.						
Start the router without the bit touching the work, and then bring it into the work, gripping the router tightly by its handles.						
Working from left to right in a counterclockwise direction, move the router along the material, keeping the pilot of the bit in contact with the edge of the material at all times.						
If the cut is not deep enough, adjust the cutting depth and repeat until the desired depth is reached.						
TOTAL POINTS						
FINAL (AVERAGED) SCORE						

EVALUATOR NOTES	STUDENT NOTES

PROCEDURE 4-5: Making and Using a Circle Cutting-Jig with the Router

Objective:	The student shall successfully demonstrate making and using a circle cutting jig with the router.
Materials Needed:	Router and wrenches
	1/2" straight router bit
	1/4" plywood or high density fiberboard for jig
	Screwdriver to fit router base plate screws
	Drill or drill press and appropriate bit
	Countersink bit
	Material to be cut into a circle
Safety Requirements:	Eye protection
	Hearing protection
References:	*Chapter 4, Portable Power Tools*
	Procedure 4-5, p. 80

Student Name: _____ Date: _____

Evaluator Name: _____ Date: _____

Ratings: 0 = Skipped Task; 1 = Attempted, but did not complete Task; 2 = Poor demonstration of Task; 3 = Fair demonstration of Task; 4 = Good demonstration of Task; 5 = Outstanding demonstration of Task.

Task	0	1	2	3	4	5
Wear proper safety equipment.						
Remove the base from the router.						
Make a new base out of 1/4" plywood or high-density fiberboard like the one diagrammed in your textbook.						
Use the factory base to determine where the screw holes for attaching the base should go.						
Drill and countersink for the attachment screws.						
Attach the new base to the router.						
Fasten the base to the material you wish to cut so that the distance from the attachment point to the router bit is the radius of the desired circle.						
Use your router like a large compass to cut the circle in a series of passes, lowering the bit a little each time.						
TOTAL POINTS						
FINAL (AVERAGED) SCORE						

EVALUATOR NOTES	STUDENT NOTES

PROCEDURE 4-6: Sanding and Smoothing a Board Face with a Belt Sander

Objective:	The student shall successfully demonstrate sanding and smoothing a board face with a belt sander.
Materials Needed:	Belt sander and belt
	Rough board to sand and smooth
	Scrap wood for stop blocks
	Clamps to hold stop blocks in place
Safety Requirements:	Eye protection
	Hearing protection
	Dust mask
	Dust bag for the sander
References:	*Chapter 4, Portable Power Tools*
	Procedure 4-6, p. 83

Student Name: _____ Date: _____

Evaluator Name: _____ Date: _____

Ratings: 0 = Skipped Task; 1 = Attempted, but did not complete Task; 2 = Poor demonstration of Task;
3 = Fair demonstration of Task; 4 = Good demonstration of Task; 5 = Outstanding demonstration of Task.

Task	0	1	2	3	4	5
Wear proper safety equipment.						
Secure the material to your bench with stop blocks to keep it from moving while sanding.						
Start the sander before it is brought into contact with the wood, and then gently lay the belt on the stock.						
Remove material quickly by setting the sander flat on the surface at a 45° angle to the grain of the wood.						
Move the sander forward immediately to prevent gouging of the wood's surface.						
Once the tool reaches the edge of the wood, pull it back until it overlaps your previous stroke by one-half the width of the belt.						
Do not allow more than half the length of the sander to run off the end or edge of the material to prevent rounding over of the edges.						
To smooth the surface of the material, use the same techniques described, but work with the sander parallel to the grain of the wood.						
TOTAL POINTS						
FINAL (AVERAGED) SCORE						

EVALUATOR NOTES	STUDENT NOTES

PROCEDURE 4-7: Joining Boards Using the Plate Joiner

Objective:	The student shall successfully demonstrate joining boards using a plate joiner.
Materials Needed:	Plate joiner
	Biscuits
	Several boards
	Glue
	Bar or pipe clamps
	Measuring tool
	Pencil
Safety Requirements:	Eye protection
	Hearing protection
References:	*Chapter 4, Portable Power Tools*
	Procedure 4-7, p. 89

Student Name: _____ Date: _____

Evaluator Name: _____ Date: _____

Ratings: 0 = Skipped Task; 1 = Attempted, but did not complete Task; 2 = Poor demonstration of Task; 3 = Fair demonstration of Task; 4 = Good demonstration of Task; 5 = Outstanding demonstration of Task.

Task	0	1	2	3	4	5
Wear proper safety equipment.						
Align the boards to be joined and mark them for easy reassembly.						
Mark the centerlines for the biscuit slots across adjacent boards, starting at least 2 inches from each end and spacing them every 6 to 8 inches.						
Adjust the fence on the plate joiner so that the slot will be cut halfway through the thickness of the boards.						
Turn the depth adjustment knob to select the appropriate depth for the biscuits.						
With a board secured to the bench with a slight overhang, align the biscuit joiner so that the center mark on the fence aligns with the mark for the slot.						
Holding the biscuit joiner with both hands, squeeze the trigger and push the biscuit joiner into the board.						
Repeat for all marks on each board.						
Reassemble the panel and stand all the boards, except the last one, on edge.						
Spread glue along the edge of each board and into each slot, inserting biscuits.						
Spread glue over the sides of each biscuit.						
Fit the boards together.						
Clamp the panel, alternating clamps on the top and bottom of the panel.						
TOTAL POINTS						
FINAL (AVERAGED) SCORE						

EVALUATOR NOTES	STUDENT NOTES

PROCEDURE 5-1: Edge Jointing a Board

Objective: The student shall successfully demonstrate edge jointing a board on the jointer.

Materials Needed: Jointer
Board

Safety Requirements: Eye protection
Hearing protection

References: *Chapter 5, Stationary Shop Tools*
Procedure 5-1, p. 109

Student Name: _____ **Date:** _____

Evaluator Name: _____ **Date:** _____

Ratings: 0 = Skipped Task; 1 = Attempted, but did not complete Task; 2 = Poor demonstration of Task; 3 = Fair demonstration of Task; 4 = Good demonstration of Task; 5 = Outstanding demonstration of Task.

Task	0	1	2	3	4	5
Wear proper safety equipment.						
Check the board to be jointed for a crown. You should run it with the crown up if it has one.						
Adjust the depth of the cut by lowering or raising the infeed table. Loosen the table lock, adjust as desired, and lock.						
Check the fence with a square to be sure it is perpendicular to the table.						
Check the way the grain runs on the board. It should run back toward the infeed table as you joint it.						
Turn on the jointer and run the board through, keeping it flush to the table and firmly against the fence. Begin with pressure on the infeed table and gradually switch it to the outfeed table, keeping your balance at all times.						
Repeat as necessary, until every part of the edge has met the knives and the board sits completely flat on the table.						
TOTAL POINTS						
FINAL (AVERAGED) SCORE						

EVALUATOR NOTES	STUDENT NOTES

PROCEDURE 5-2: Ripping on the Table Saw

Objective: The student shall successfully demonstrate ripping on the table saw.
Materials Needed: Table saw
Board
Safety Requirements: Eye protection
Hearing protection
Push stick
References: *Chapter 5, Stationary Shop Tools*
Procedure 5-2, p. 122

Student Name: _____ **Date:** _____

Evaluator Name: _____ **Date:** _____

Ratings: 0 = Skipped Task; 1 = Attempted, but did not complete Task; 2 = Poor demonstration of Task; 3 = Fair demonstration of Task; 4 = Good demonstration of Task; 5 = Outstanding demonstration of Task.

Task	0	1	2	3	4	5
Wear proper safety equipment.						
Unlock and slide the rip fence for the desired width of cut.						
Lock the fence and check the distance from the fence to the blade to ensure that the distance is correct. Adjust if necessary.						
Unlock the locking hub on the blade height adjustment handwheel, and set the blade height so that one full tooth and part of an adjacent gullet will protrude above the wood.						
Turn on the saw, and butt the edge of the board to be ripped against the fence.						
Standing so that you are not directly in line with the blade, push the board through the blade; keep it down on the table and against the fence.						
Finish the cut using a push stick if the piece is narrower than 6". Push the piece all the way past the blade.						
TOTAL POINTS						
FINAL (AVERAGED) SCORE						

EVALUATOR NOTES	STUDENT NOTES

PROCEDURE 5-3: Crosscutting Several Boards to the Same Length

Objective:	The student shall successfully demonstrate crosscutting several boards to the same length on the table saw.
Materials Needed:	Table saw
	Miter gauge
	Straight board for miter gauge
	Screws
	Screwdriver
	Framing square
	Square
	Pencil
	Scrap to use as a stop block
	Clamp
	Several boards to cut
Safety Requirements:	Eye protection
	Hearing protection
References:	*Chapter 5, Stationary Shop Tools*
	Procedure 5-3, p. 124

Student Name: _____   Date: _____

Evaluator Name: _____   Date: _____

Ratings: 0 = Skipped Task; 1 = Attempted, but did not complete Task; 2 = Poor demonstration of Task; 3 = Fair demonstration of Task; 4 = Good demonstration of Task; 5 = Outstanding demonstration of Task.

Task	0	1	2	3	4	5
Wear proper safety equipment.						
Attach the face of a straight board to the miter gauge with screws to serve as an auxiliary fence.						
Use a framing square to check that the miter gauge is at 90° to the saw blade.						
On one board, measure and mark the length of the pieces to be cut.						
Slide the miter gauge up to the blade, and adjust the board so that it will be cut on the waste side of the line.						
Clamp a stop block on the auxiliary fence to limit the length of the cut.						
Butt the piece to be cut against the stop block.						
Turn on the saw, and holding the piece against the miter gauge, slide the miter gauge forward and past the blade.						
Check that the cut was made in the proper place, and adjust the stop block if necessary.						
Make successive cuts of the same length by butting each piece against the stop and pushing it past the blade.						
TOTAL POINTS						
FINAL (AVERAGED) SCORE						

EVALUATOR NOTES	STUDENT NOTES

236

PROCEDURE 5-4: Changing a Band Saw Blade

Objective: The student shall successfully demonstrate changing a blade on the band saw.
Materials Needed: Band saw
Band saw blade
Safety Requirements: Optional: gloves
References: *Chapter 5, Stationary Shop Tools*
Procedure 5-4, p. 140

Student Name: _____   **Date:** _____

Evaluator Name: _____   **Date:** _____

Ratings: 0 = Skipped Task; 1 = Attempted, but did not complete Task; 2 = Poor demonstration of Task;
3 = Fair demonstration of Task; 4 = Good demonstration of Task; 5 = Outstanding demonstration of Task.

Task	0	1	2	3	4	5
Disconnect the power to the saw.						
Swing aside or remove the upper and lower wheel guards.						
Release the tension on the blade by lowering the top wheel.						
Back off the guide components and the thrust bearings both above and below the table.						
Remove the table insert and table leveling pin and slide the old blade off the wheels and through the slot in the table.						
Slide the new blade onto the wheels, making sure the teeth are facing down and toward the front of the saw.						
Reset the tension for the size blade installed using the guide on the back of the saw.						
Replace the table insert and the table leveling pin.						
Adjust the blade guide components and thrust bearings both above and below the table.						
Spin the blade by hand to make sure it is tracking correctly.						
Replace the wheel guards, reconnect the power, and turn the saw on and then off, checking to see that the blade is tracking correctly.						
TOTAL POINTS						
FINAL (AVERAGED) SCORE						

EVALUATOR NOTES	STUDENT NOTES

PROCEDURE 5-5: Drilling to Exact Depth

Objective: The student shall successfully demonstrate drilling to exact depth on the drill press.

Materials Needed: Drill press
Drill bit
Board
Measuring tool
Pencil

Safety Requirements: Eye protection

References: *Chapter 5, Stationary Shop Tools*
Procedure 5-5, p. 146

Student Name: _____   Date: _____

Evaluator Name: _____   Date: _____

Ratings: 0 = Skipped Task; 1 = Attempted, but did not complete Task; 2 = Poor demonstration of Task; 3 = Fair demonstration of Task; 4 = Good demonstration of Task; 5 = Outstanding demonstration of Task.

Task	0	1	2	3	4	5
Wear proper safety equipment.						
Mark the depth to which you wish to drill on the side of the piece to be drilled.						
Set the marked piece on the table of the drill press.						
Pull on the quill feed lever to bring the bit down beside the piece even with the mark you made.						
Set the depth stop.						
Align the bit where you wish to drill the hole, turn the drill press on, and bring down the feed lever to bring down the drill; it will stop at the preset depth.						
TOTAL POINTS						
FINAL (AVERAGED) SCORE						

239

EVALUATOR NOTES	STUDENT NOTES

PROCEDURE 6-1: Clamping a Wide Panel

Objective: The student shall successfully demonstrate clamping a wide panel.

Materials Needed: Boards for panel
Jointer
Table saw
Bar or pipe clamps
Glue
Pencil

Safety Requirements: Eye protection
Hearing protection

References: *Chapter 6, Clamps*
Procedure 6-1, p. 190

Student Name: _____ **Date:** _____

Evaluator Name: _____ **Date:** _____

Ratings: 0 = Skipped Task; 1 = Attempted, but did not complete Task; 2 = Poor demonstration of Task;
3 = Fair demonstration of Task; 4 = Good demonstration of Task; 5 = Outstanding demonstration of Task.

Task	0	1	2	3	4	5
Wear proper safety equipment.						
Rip and joint boards to be used in the panel. Cut 1 or 2 inches longer than the finished length of the panel, and allow extra width as well.						
Arrange boards in a pleasing pattern, alternating the growth rings, and mark for easy reassembly.						
Place bar or pipe clamps on a flat surface and put the boards on it. There should be clamps near each end, and spaced 10 to 16 inches apart.						
Turn all the boards but the last one on edge and apply glue, spreading it evenly.						
Lay the boards back down on the clamps and snug up the clamps, tapping the boards with a mallet if necessary to get them properly aligned. To avoid marring the edges of the panel, use scrap material on either side of the panel.						
Place clamps over the top of the panel, spacing them between the bottom clamps; snug them up.						
Go through and tighten all the clamps. They should all be snug.						
If any of the joints are not flush, use hand screws on either end of the joints or cauls and C-clamps to bring the joints flush.						
TOTAL POINTS						
FINAL (AVERAGED) SCORE						

EVALUATOR NOTES	STUDENT NOTES

PROCEDURE 6-2: Squaring Up an Assembly

Objective: The student shall successfully demonstrate squaring up an assembly.

Materials Needed: A glued frame assembly
Measuring tool
Clamp
Clamp blocks (optional)

Safety Requirements: None

References: *Chapter 6, Clamps*
Procedure 6-2, p. 192

Student Name: _____ Date: _____

Evaluator Name: _____ Date: _____

Ratings: 0 = Skipped Task; 1 = Attempted, but did not complete Task; 2 = Poor demonstration of Task;
3 = Fair demonstration of Task; 4 = Good demonstration of Task; 5 = Outstanding demonstration of Task.

Task	0	1	2	3	4	5
Check a glued assembly to see if it is square by measuring diagonals. If they are the same, it is square.						
If the assembly is out of square, place a clamp across the long diagonal. Corners can be protected by making and using the corner blocks diagrammed in your textbook.						
Slowly tighten the clamp, and check the diagonal measurements again. Continue until the assembly is square.						
TOTAL POINTS						
FINAL (AVERAGED) SCORE						

EVALUATOR NOTES	STUDENT NOTES

PROCEDURE 6-3: Keeping Assemblies Flat

Objective: The student shall successfully demonstrate checking an assembly for flat and keeping an assembly flat.

Materials Needed: A glued assembly such as a box or drawer
Winding sticks
Clamps or weights

Safety Requirements: None

References: *Chapter 6, Clamps*
Procedure 6-3, p. 199

Student Name: _____ Date: _____

Evaluator Name: _____ Date: _____

Ratings: 0 = Skipped Task; 1 = Attempted, but did not complete Task; 2 = Poor demonstration of Task;
3 = Fair demonstration of Task; 4 = Good demonstration of Task; 5 = Outstanding demonstration of Task.

Task	0	1	2	3	4	5
Place the clamped up assembly on a flat surface.						
Check to see whether the assembly is flat; it should be touching the surface at all points.						
Also check for flatness by placing two equal-sized sticks (winding sticks) on the top of the assembly and sighting across them horizontally. If the sticks appear parallel, the assembly is flat.						
If the assembly is not flat, clamp it down to the table or weight it down until it is flat. Leave the assembly clamped or weighted down until the glue is dry.						
TOTAL POINTS						
FINAL (AVERAGED) SCORE						

EVALUATOR NOTES	STUDENT NOTES

PROCEDURE 7-1: Driving a Nail

Objective: The student shall successfully demonstrate driving and setting a finish nail.

Materials Needed: Two boards
Finish nail
Hammer
Nail set appropriate to the size of the finish nail

Safety Requirements: Eye protection

References: *Chapter 7, Fasteners*
Procedure 7-1, p. 212

Student Name: _____ Date: _____

Evaluator Name: _____ Date: _____

Ratings: 0 = Skipped Task; 1 = Attempted, but did not complete Task; 2 = Poor demonstration of Task; 3 = Fair demonstration of Task; 4 = Good demonstration of Task; 5 = Outstanding demonstration of Task.

Task	0	1	2	3	4	5
Wear proper safety equipment.						
Hold the nail between your thumb and forefinger.						
Tap the nail with your hammer to get it started.						
Pull back your hand, keep your eye on the nail head, and use your wrist to swing the hammer.						
Take as many swings as needed to drive the nail close to the surface and stop.						
Place a nail set on the nail head and strike the nail set to drive the nail beneath the surface of the wood.						
TOTAL POINTS						
FINAL (AVERAGED) SCORE						

EVALUATOR NOTES	STUDENT NOTES

PROCEDURE 7-2: Installing a Countersunk Screw in Hardwood

Objective: The student shall successfully demonstrate installing a countersunk screw in hardwood.

Materials Needed: Two boards
Screw of the appropriate length
Drill
Properly sized bits for shank and pilot holes
Countersink bit
Screwdriver

Safety Requirements: Eye protection

References: *Chapter 7, Fasteners*
Procedure 7-2, p. 215

Student Name: _____ **Date:** _____

Evaluator Name: _____ **Date:** _____

Ratings: 0 = Skipped Task; 1 = Attempted, but did not complete Task; 2 = Poor demonstration of Task;
3 = Fair demonstration of Task; 4 = Good demonstration of Task; 5 = Outstanding demonstration of Task.

Task	0	1	2	3	4	5
Wear proper safety equipment.						
Select a screw that is long enough so that all of the threaded portion will go into the bottom board. Approximately two-thirds of the screw should be in the lower piece, as shown here. Use a thinner gauge screw for thinner woods and a heavier gauge for thicker materials.						
Hold the two pieces to be attached to each other together and drill a shank hole equal to or slightly larger than the diameter of the screw; drill all the way through the first piece and just into the second. The screw should slip through this hole without pressure.						
See the chart in your textbook to select the proper sized bit, and drill a pilot hole the length of the screw.						
Install a countersink bit in your drill and make the countersink.						
Check the size of the countersink with the head of the screw by inverting the screw.						
Drive the screw with the properly sized driver. When set, it should be flush with the surface of the work.						
TOTAL POINTS						
FINAL (AVERAGED) SCORE						

EVALUATOR NOTES	STUDENT NOTES

PROCEDURE 7-3: Making and Installing Plugs

Objective: The student shall successfully demonstrate making and installing plugs.

Materials Needed: Board with holes the same size as the plug cutter
Plug cutter
Scrap stock
Drill press
Band saw
Mallet
Glue
Block plane
Sandpaper

Safety Requirements: Eye protection
Hearing protection

References: *Chapter 7, Fasteners*
Procedure 7-3, p. 217

Student Name: _____ Date: _____

Evaluator Name: _____ Date: _____

Ratings: 0 = Skipped Task; 1 = Attempted, but did not complete Task; 2 = Poor demonstration of Task; 3 = Fair demonstration of Task; 4 = Good demonstration of Task; 5 = Outstanding demonstration of Task.

Task	0	1	2	3	4	5
Wear proper safety equipment.						
Install a plug-cutting bit of the proper size for the holes you want to plug in the drill press.						
Select the piece you will use to make the plugs and place it on the drill press table. Bring down the plug cutter, and set the depth stop on the drill press so that you do not cut all the way through the material.						
Turn on the drill press and bore the plugs.						
Take the material to the band saw and make a resaw cut to free the plugs.						
Install the plugs in the holes with glue and tap them in place. Orient the grain direction of each plug so that it matches the grain direction of the workpiece; this makes it less noticeable. The plugs will stand above the workpiece.						
When the glue is dry, plane and sand the plugs flush with the surrounding material.						
TOTAL POINTS						
FINAL (AVERAGED) SCORE						

EVALUATOR NOTES	STUDENT NOTES

PROCEDURE 8-1: Using Adhesives

Objective: The student shall successfully demonstrate using adhesives.
Materials Needed: Unglued assembly
                          Glue
                          Glue brush (optional)
                          Clamps
Safety Requirements: None
References: *Chapter 8, Adhesives*
                          Procedure 8-1, p. 243

Student Name: _____   Date: _____

Evaluator Name: _____   Date: _____

Ratings: 0 = Skipped Task; 1 = Attempted, but did not complete Task; 2 = Poor demonstration of Task; 3 = Fair demonstration of Task; 4 = Good demonstration of Task; 5 = Outstanding demonstration of Task.

Task	0	1	2	3	4	5
Fit your assembly together before gluing. This is called *dry assembly*. Get out the clamps you will need and actually clamp up the project to ensure that it will go together without any problem. If you discover a poorly fitting joint, fix it before proceeding.						
Release the clamps and leave them handy. Wipe any dust or debris off the pieces.						
Apply glue in thin, even coats to the surfaces to be joined. Spread it evenly with a small brush, scrap of wood, or your finger. You are working on what is called an *open assembly*.						
Clamp the assembly. It is now referred to as a *closed assembly*. With the exception of epoxies, contact cement, and cyanoacrylate glue, wood glues must cure under pressure to form a strong bond. Small beads of glue should ooze from the joint; this tells you that sufficient glue was used.						
TOTAL POINTS						
FINAL (AVERAGED) SCORE						

EVALUATOR NOTES	STUDENT NOTES

PROCEDURE 9-1: Calculating Board Footage

Objective: The student shall successfully demonstrate calculating board footage and price of a given species of hardwood.

Materials Needed: Calculator
Two or more hardwood boards of the same species, real or imaginary

Safety Requirements: None

References: *Chapter 9, Wood*
Procedure 9-1, p. 268

Student Name: _____ Date: _____

Evaluator Name: _____ Date: _____

Ratings: 0 = Skipped Task; 1 = Attempted, but did not complete Task; 2 = Poor demonstration of Task; 3 = Fair demonstration of Task; 4 = Good demonstration of Task; 5 = Outstanding demonstration of Task.

Task	0	1	2	3	4	5
Measure the thickness, width, and length of one board in inches.						
Multiply the thickness, width, and length of the board together.						
Divide the sum obtained in the previous step by 144.						
Repeat for each additional board.						
Add together the number determined for each board to determine the total board footage.						
Round off to the nearest whole board foot.						
Find out what the board foot price for that type of hardwood.						
Multiply the total board feet by the price to determine the pretax cost of your hardwood.						
TOTAL POINTS						
FINAL (AVERAGED) SCORE						

EVALUATOR NOTES	STUDENT NOTES

PROCEDURE 10-1: Applying Adhesive-Backed Edge Banding

Objective: The student shall successfully demonstrate applying adhesive-backed edge banding.

Materials Needed: Panel product material
Adhesive-backed edge banding
Scissors or knife
Edge banding iron or household iron
Roller or block of wood
Edge trimmer
Sandpaper

Safety Requirements: None

References: *Chapter 10, Panel Products*
Procedure 10-1, p. 293

Student Name: _____ Date: _____

Evaluator Name: _____ Date: _____

Ratings: 0 = Skipped Task; 1 = Attempted, but did not complete Task; 2 = Poor demonstration of Task; 3 = Fair demonstration of Task; 4 = Good demonstration of Task; 5 = Outstanding demonstration of Task.

Task	0	1	2	3	4	5
Using scissors or a knife, cut a length of edge banding slightly longer than needed to cover the edge of your panel product.						
Secure the piece to be banded in a vise or in some other manner so that the edge to be covered is facing up.						
Place the edge banding on the edge of the panel product so that it is centered, with the ends overhanging slightly.						
Holding the edge banding in place with one hand, apply heat to it with a household iron or an edge banding iron.						
Move the heating device back and forth over the edge banding without stopping in any one spot. This activates the glue on the bottom of the edge banding.						
Once the edge banding begins to stick, take away the heat and use a roller or a block of wood to vigorously rub the banding in place.						
Trim the projecting ends, and then use an edge trimmer to trim the veneer flush with the sides of the panel.						
Lightly sand the edges of the veneer by hand to blend it in with the panel surfaces.						
TOTAL POINTS						
FINAL (AVERAGED) SCORE						

EVALUATOR NOTES	STUDENT NOTES

PROCEDURE 15-1: Making a Rabbet Joint on the Table Saw

Objective: The student shall successfully demonstrate making a rabbet joint on the table saw.

Materials Needed: Table saw
Dado blade
Sacrificial fence and clamps for attaching it to rip fence
Miter gauge
Scrap material for test cuts
Board(s) to rabbet and board(s) to fit into rabbet

Safety Requirements: Eye protection
Hearing protection

References: *Chapter 15, Case Joints*
Procedure 15-1, p. 378

Student Name: _____   **Date:** _____

Evaluator Name: _____   **Date:** _____

Ratings: 0 = Skipped Task; 1 = Attempted, but did not complete Task; 2 = Poor demonstration of Task; 3 = Fair demonstration of Task; 4 = Good demonstration of Task; 5 = Outstanding demonstration of Task.

Task	0	1	2	3	4	5
Wear proper safety equipment.						
Install a dado blade on the saw; set it up for the width of the rabbet you want to make or a bit wider.						
Clamp a sacrificial fence to the table saw fence, and bring the fence over next to the blade.						
Raise the dado blade to the height desired. If the blade width is wider than the desired width of the rabbet, the blade needs to be raised into the sacrificial fence with the saw turned on.						
With the fence in the correct position and the blade at the proper height, run a test piece with the thickness of the material you will be working with; support it with the miter gauge.						
Check the test piece against the piece it will butt into to check the fit.						
Adjust the fence position and blade height as needed, and keep running test pieces until the proper size rabbet has been achieved.						
Run all pieces to be rabbeted.						
TOTAL POINTS						
FINAL (AVERAGED) SCORE						

EVALUATOR NOTES	STUDENT NOTES

PROCEDURE 15-2: Machining a Dado-and-Rabbet Joint

Objective: The student shall successfully demonstrate machining a dado-and-rabbet joint on the table saw.

Materials Needed: Table saw
Dado blade
Miter gauge
Sacrificial fence and clamps (for rabbeting operation)
Scrap material for test cuts
Board(s) to dado and board(s) to rabbet

Safety Requirements: Eye protection
Hearing protection

References: *Chapter 15, Case Joints*
Procedure 15-2, p. 380

Student Name: _____ Date: _____

Evaluator Name: _____ Date: _____

Ratings: 0 = Skipped Task; 1 = Attempted, but did not complete Task; 2 = Poor demonstration of Task;
3 = Fair demonstration of Task; 4 = Good demonstration of Task; 5 = Outstanding demonstration of Task.

Task	0	1	2	3	4	5
Wear proper safety equipment.						
Install the dado blade on the table saw with its width set for the dado width desired.						
Set the blade height for the depth of the dado desired.						
Slide the rip fence over to act as a stop for the dado joint you will make. The distance from the fence to the far side of the blade should be the thickness of the piece to be rabbeted.						
Using the miter gauge to support the workpiece, make a test dado.						
Check the depth and placement of the dado, and adjust the blade height and fence position until you are satisfied.						
Mill the dado in your finish stock.						
Next the rabbet is cut. Follow the steps outlined in Procedure 15-1, making test cuts until you have a rabbet that fits snugly into the dado.						
TOTAL POINTS						
FINAL (AVERAGED) SCORE						

EVALUATOR NOTES	STUDENT NOTES

PROCEDURE 15-3: Making a Splined Miter Joint

Objective: The student shall successfully demonstrate making a splined miter joint.
Materials Needed: Table saw
Miter saw (optional)
Miter gauge
Scrap material for test cuts
Two boards to miter
1/8" hardboard for spline
Clamps
Safety Requirements: Eye protection
Hearing protection
References: *Chapter 15, Case Joints*
Procedure 15-3, p. 385

Student Name: _____ **Date:** _____

Evaluator Name: _____ **Date:** _____

Ratings: 0 = Skipped Task; 1 = Attempted, but did not complete Task; 2 = Poor demonstration of Task;
3 = Fair demonstration of Task; 4 = Good demonstration of Task; 5 = Outstanding demonstration of Task.

Task	0	1	2	3	4	5
Wear proper safety equipment.						
Cut the mitered corners that will be joined using the table saw or miter saw.						
Tilt the table saw blade to 45°. The blade should tilt away from the fence. If necessary, move the fence to the other side of blade.						
Using one of your mitered pieces as a guide, set the fence so that the blade will be cutting into the inside third of the miter.						
Adjust the blade height so that it will cut approximately halfway through the mitered stock.						
With the point of the miter against the fence and the workpiece supported by the miter gauge, make a cut in a test piece.						
Adjust the fence position and blade height as needed, and continue to test the setup until you are satisfied.						
Run your finish stock.						
Bring the table saw blade back to its normal position at 90° to the table, and rip a strip of eighth-inch hardboard that is just slightly less than twice the depth of the mitered slot you just cut.						
Check the fit of the spline.						
Cut the spline to length.						
Apply glue to the spline, insert it into the mitered grooves of the mating corner, and clamp.						
TOTAL POINTS						
FINAL (AVERAGED) SCORE						

EVALUATOR NOTES	STUDENT NOTES

PROCEDURE 15-4: Cutting a Dovetail Joint by Hand

Objective:	The student shall successfully demonstrate cutting a dovetail joint by hand.
Materials Needed:	Four boards of the same width
	Miter saw to cut boards to length
	Marking knife
	Square
	Marking gauge (optional)
	Pencil
	Bevel square or dovetail gauge
	Dovetail or dozuki saw
	Chisels
	Mallet
Safety Requirements:	Eye protection
	Hearing protection (if using miter saw)
References:	*Chapter 15, Case Joints*
	Procedure 15-4, p. 392

Student Name: _____ Date: _____

Evaluator Name: _____ Date: _____

Ratings: 0 = Skipped Task; 1 = Attempted, but did not complete Task; 2 = Poor demonstration of Task; 3 = Fair demonstration of Task; 4 = Good demonstration of Task; 5 = Outstanding demonstration of Task.

Task	0	1	2	3	4	5
Wear proper safety equipment.						
Select good, straight-grained stock; it is easiest to work. You will need four pieces that are of the same width. Cut the ends square.						
Mark the shoulder on both ends of all four pieces. The shoulder is set over from the end of the board the distance of the board's thickness plus a bit. These are best scribed with a marking gauge or a sharp marking knife and square.						
Run a sharp pencil along the scribed line to make it more visible.						
Using your bevel square set to the proper angle or a dovetail gauge, mark out the tails on both ends of the two tail boards, sizing and spacing them according to your preference.						
Secure one of the tail boards in a vise, and carefully cut down each sloped line. A fine-toothed saw such as a dovetail or dozuki saw should be used. Cut all the tails you marked out on both ends of the two boards.						
After all the angled cuts for the tails have been made, secure the tailboard with the flat face down to the bench, and with an appropriately sized chisel, chop down into the scribed shoulder line; then come in horizontally from the end, and split out the waste. Do this a bit at a time, until you are about halfway through, and then turn the board over and finish from the other side.						

Continued

Task	0	1	2	3	4	5
After the tails have been cut out, the pin boards are laid out. Lay out your box on the bench as it will go together, and mark each corner.						
The pins for each corner are laid out from the tails. Secure the pin board in your vise and trace the tails on to it. You will notice that where the slope on the tails was on the face of the board, the pins are sloped on the ends.						
With the pins laid out across the ends, lines are squared down to the shoulder line.						
Now these are cut and cleaned out in the same manner as the tails. Be careful to cut to the waste side of the line.						
When the pins are cut, test the joint for fit. Adjust fit as needed.						
TOTAL POINTS						
FINAL (AVERAGED) SCORE						

EVALUATOR NOTES	STUDENT NOTES

PROCEDURE 16-1: Creating a Pocket Joint

Objective:	The student shall successfully demonstrate creating a pocket joint.
Materials Needed:	Two stiles
	Two (or three) rails
	Tools to size stiles and rails
	Pocket hole jig and bit
	Pocket hole screws
	Long driver
Safety Requirements:	Eye protection
	Hearing protection
References:	*Chapter 16, Frame Joints*
	Procedure 16-1, p. 408

Student Name: _____ **Date:** _____

Evaluator Name: _____ **Date:** _____

Ratings: 0 = Skipped Task; 1 = Attempted, but did not complete Task; 2 = Poor demonstration of Task; 3 = Fair demonstration of Task; 4 = Good demonstration of Task; 5 = Outstanding demonstration of Task.

Task	0	1	2	3	4	5
Wear proper safety equipment.						
Cut the frame pieces to size.						
Place a rail into the jig, or attach the jig to that piece, depending on the type of jig you have.						
With the special stepped drill bit that came with the jig, drill two holes through the rail. This bit drills a pilot hole and a recess for the screw head, as well as creating room for the drill bit that drives the screw.						
Repeat at both ends of each rail.						
Place two pieces to be joined on a flat surface and clamp.						
Drive self-tapping pan-head pocket-hole screws through the rail into the stile.						
Repeat for all stile/rail intersections.						
TOTAL POINTS						
FINAL (AVERAGED) SCORE						

EVALUATOR NOTES	STUDENT NOTES

PROCEDURE 16-2: Machining a Half-Lap Joint

Objective:	The student shall successfully demonstrate machining a half-lap joint.
Materials Needed:	Two stiles
	Two (or three) rails
	Tools to size stiles and rails
	Scrap the same thickness as the stiles and rails for tests
	Table saw
	Dado blade
	Miter gauge
Safety Requirements:	Eye protection
	Hearing protection
References:	*Chapter 16, Frame Joints*
	Procedure 16-2, p. 416

Student Name: _____   Date: _____

Evaluator Name: _____   Date: _____

Ratings: 0 = Skipped Task; 1 = Attempted, but did not complete Task; 2 = Poor demonstration of Task;
3 = Fair demonstration of Task; 4 = Good demonstration of Task; 5 = Outstanding demonstration of Task.

Task	0	1	2	3	4	5
Wear proper safety equipment.						
Cut the pieces to the sizes required.						
Mount a dado blade set for its greatest width on the table saw, and raise it so the amount of blade protruding from the table is equal to one-half the thickness of the material being machined.						
Lay out the joint carefully on a test piece. If the width of all the component pieces is the same, this will only need to be done on one piece.						
Adjust the rip fence so that the far side of the blade will cut just inside your layout line, when the board is butted against the fence.						
Support the test piece with the miter gauge and push it over the spinning dado blade. After making one pass, move the piece to the left and run it again. Continue to do this until you reach the end of the piece.						
Repeat the previous step with a second test piece, and then check the fit of the two pieces. They should be flush with each other and meet squarely at the ends.						
Adjust the fence if the two pieces do not match at the ends, and adjust the blade height if they are not flush. Continue to run test pieces until you are satisfied with the fit.						
Run your finish pieces.						
TOTAL POINTS						
FINAL (AVERAGED) SCORE						

269

EVALUATOR NOTES	STUDENT NOTES

PROCEDURE 16-3: Making a Mitered Half-Lap Joint

Objective: The student shall successfully demonstrate making a mitered half-lap joint.
Materials Needed: Two boards of equal width
Table saw
Dado blade
Miter gauge
Miter saw
Safety Requirements: Eye protection
Hearing protection
References: *Chapter 16, Frame Joints*
Procedure 16-3, p. 418

Student Name: _____   Date: _____

Evaluator Name: _____   Date: _____

Ratings: 0 = Skipped Task; 1 = Attempted, but did not complete Task; 2 = Poor demonstration of Task;
3 = Fair demonstration of Task; 4 = Good demonstration of Task; 5 = Outstanding demonstration of Task.

Task	0	1	2	3	4	5
Wear proper safety equipment.						
Set up the table saw as if you were going to make a half-lap joint. Refer to Procedure 16-2.						
Take one of the two pieces that will form the corner and mill a half lap on it.						
Take the piece that you just machined to the miter saw and cut a miter on the same end that you created the half lap.						
Using the piece with the half lap as a guide, set the depth stop on the miter saw so that the saw will only cut down as far as that half lap.						
Take the other piece that will form the corner to the miter saw; swing the saw so that it will cut the mating miter to the first piece. Make the miter from the corner.						
Remove the rest of the material by moving the piece and taking successive cuts on the miter saw.						
Use a chisel to remove any ridges left by the miter saw, and check the fit of the joint. If necessary, adjust the depth stop on the miter saw until you get the desired fit.						
TOTAL POINTS						
FINAL (AVERAGED) SCORE						

EVALUATOR NOTES	STUDENT NOTES

PROCEDURE 16-4: Machining a Groove-and-Stub Tenon Joint

Objective: The student shall successfully demonstrate machining a groove-and-stub tenon joint.

Materials Needed: Two stiles
Two (or three) rails
Material for test cuts
Table saw
Rip or combination blade
Dado blade
Sacrificial fence
Miter gauge

Safety Requirements: Eye protection
Hearing protection

References: *Chapter 16, Frame Joints*
Procedure 16-4, p. 420

Student Name: _____ Date: _____

Evaluator Name: _____ Date: _____

Ratings: 0 = Skipped Task; 1 = Attempted, but did not complete Task; 2 = Poor demonstration of Task; 3 = Fair demonstration of Task; 4 = Good demonstration of Task; 5 = Outstanding demonstration of Task.

Task	0	1	2	3	4	5
Wear proper safety equipment.						
Determine the sizes of the component pieces based on your needs.						
Make a centered groove on a test piece that is the same thickness as your finish pieces by raising the blade of the table saw to the desired groove depth. Mark the center of the piece's thickness, and set the rip fence so that the blade will cut just on the side of your mark.						
Run the piece; flip it so that the opposite face is against the fence, and run it again.						
Check the depth of the groove and adjust the blade height as needed.						
Now check the width of the groove. Do this by testing it with the material that will be used for the panel. The fit should be snug, but not overly tight. If the fit is too loose, move the fence slightly toward the blade. If too tight, move it slightly away from the blade.						
Once you are satisfied, run the grooves on the inside of the two stiles and the top and bottom rails. If there is a middle rail, as in the example, it gets grooves on both edges.						
Cut your final correct test piece in half. You will use it for the next setup.						

Continued

Task	0	1	2	3	4	5
Put a dado blade on the table saw, setting its width so that it is the same as the depth of the groove. Use one of the test pieces as a gauge to set the height of the blade. The blade height is set so that it is just shy of cutting into the groove.						
Clamp a sacrificial fence to the rip fence, and bring the fence right over next to the blade.						
Using one of the test pieces supported by the miter gauge, and with its end against the rip fence, make a cut. Then flip it over and make a second cut.						
Check the fit of the resulting stub tenon in the groove of the other test piece. The tenon should bottom out in the groove, and the joint should pull tight. If the tenon is too long, move the fence in toward the blade. If it is too short, move the fence away from the blade. The fit of the tenon in the groove should be snug. If it is too tight, raise the blade a bit. Remember that you are taking equal amounts off each side. If the fit is too loose, lower the blade.						
Keep testing until you have a good fit, and then machine all the pieces that require tenons.						
TOTAL POINTS						
FINAL (AVERAGED) SCORE						

EVALUATOR NOTES	STUDENT NOTES

PROCEDURE 16-5: Creating a Blind Mortise-and-Tenon Joint

Objective:	The student shall successfully demonstrate creating a blind mortise-and-tenon joint.
Materials Needed:	Two stiles
	Two (or three) rails
	Square
	Pencil or marking knife
	Mortising machine and chisel or drill press with mortising attachment
	Table saw
	Miter gauge
	Tenoning jig
	Band saw and fence or handsaw
Safety Requirements:	Eye protection
	Hearing protection
References:	***Chapter 16, Frame Joints***
	Procedure 16-5, p. 427

Student Name: _____   Date: _____

Evaluator Name: _____   Date: _____

Ratings: 0 = Skipped Task; 1 = Attempted, but did not complete Task; 2 = Poor demonstration of Task;
3 = Fair demonstration of Task; 4 = Good demonstration of Task; 5 = Outstanding demonstration of Task.

Task	0	1	2	3	4	5
Wear proper safety equipment.						
Determine the sizes of the component pieces based on your needs.						
Refer to your plans. Carefully lay out one mortise. Mark the depth of the mortise on the end of this piece. The depth of the mortise is the length of the tenon plus 1/16".						
Transfer the beginning and end marks of the mortise from the piece you just laid out to all the other places that will get mortises.						
Install the proper size mortising chisel in the mortising machine.						
Clamp the piece with the completely laid-out mortise into the mortising machine in such a way that you will be able to set the table height for the depth of cut and do so.						
Re-clamp your piece and bring it under the mortising chisel. Adjust the table so that the chisel will come down in the proper spot.						
Make the mortise, making cuts at the beginning and end of the laid-out mortise, and then take overlapping passes to remove the rest of the waste.						
Repeat for other mortises.						
Set the mortised pieces aside, and carefully lay out one of the tenons.						
At the table saw, using the piece you just marked out as a guide, set the blade height so that it will cut just to your line.						
Slide the rip fence over so that your cut will be made just inside the layout line.						

Continued

Task	0	1	2	3	4	5
Supporting the piece with the miter gauge and butting it against the fence, make your first cut.						
Flip the piece over and make a second cut.						
Repeat for all other tenons.						
Without moving the rip fence, place your piece on edge and raise the blade height so that it will cut on the layout line.						
Again butting the piece against the fence and supporting it with the miter gauge, make a cut and then flip the piece over and make a second cut. Repeat for all tenons. You have now defined the shoulders of each tenon.						
Slide the rip fence out of the way. Set your piece on end on the saw table, and raise the blade until it will cut just into the kerf created earlier.						
Clamp your piece into the tenoning jig, and adjust the jig so that you will cut just inside the layout line on the edge of your piece.						
Turn on the saw and move the jig past the blade.						
Remove your piece, turn it around, and run it again.						
Check the fit of the tenon in a mortise, and adjust the jig as needed.						
Repeat for each tenon. You have now defined the cheeks of the tenons.						
Square the two lines that remain on the end of your piece up onto the tenon cheeks.						
Carefully cut down these lines to the shoulder of the tenon, using a handsaw or band saw. If you use the band saw, you can set the fence to help guide the cut, and you will only have to lay out one piece.						
Repeat for all tenons.						
Clean out any debris in each mortise, and use a chisel or rabbeting plane to do any fine-tuning of the tenons.						
Make a final check for fit prior to gluing.						
Glue and clamp the joints.						
TOTAL POINTS						
FINAL (AVERAGED) SCORE						

EVALUATOR NOTES	STUDENT NOTES

PROCEDURE 18-1: Making a Dado-and-Rabbet Housing

Objective: The student shall successfully demonstrate making a dado-and-rabbet housing.

.Materials Needed: Two boards of the same width
Dado jig as described in Chapter 18 or straightedge
Clamps
Square
Pencil
Router
Straight bit

Safety Requirements: Eye protection
Hearing protection

References: *Chapter 18, Housed Joints*
Procedure 18-1, p. 472

Student Name: _____ Date: _____

Evaluator Name: _____ Date: _____

Ratings: 0 = Skipped Task; 1 = Attempted, but did not complete Task; 2 = Poor demonstration of Task; 3 = Fair demonstration of Task; 4 = Good demonstration of Task; 5 = Outstanding demonstration of Task.

Task	0	1	2	3	4	5
Wear proper safety equipment.						
Mark out the position of the dado on one board.						
Fit the appropriate size straight bit in a router for the width of dado desired, and set it for the depth required.						
If using a jig, like the one described in Chapter 18, clamp it to the work with the 1/4" edge lined up with your layout mark. If a straightedge is clamped to the work, you need to figure out the distance from the edge of the router base to the bit and offset the straightedge by this amount.						
Turn on the router and machine the dado; keep the router base tight against the jig or straightedge.						
Lay out the rabbet on the second board, and adjust the depth of cut as needed so that the rabbet will fit into the dado.						
Set and clamp the jig or straightedge so that the rabbet will be cut in the proper spot.						
Mill the rabbet.						
Fit the rabbet into the dado.						
TOTAL POINTS						
FINAL (AVERAGED) SCORE						

EVALUATOR NOTES	STUDENT NOTES

PROCEDURE 19-1: Milling a Tongue-and-Groove Joint for a Wide Panel

Objective: The student shall successfully demonstrate milling a tongue-and-groove joint for a wide panel.

.Materials Needed: Sufficient 3/4" stock for the size of panel desired
Table saw
Rip or combination blade
Dado blade
Pencil
Glue
Clamps

Safety Requirements: Eye protection
Hearing protection

References: *Chapter 19, Making Wide Panels, Thick Blanks, and Corner Joints*
Procedure 19-1, p. 486

Student Name: _____ **Date:** _____

Evaluator Name: _____ **Date:** _____

Ratings: 0 = Skipped Task; 1 = Attempted, but did not complete Task; 2 = Poor demonstration of Task; 3 = Fair demonstration of Task; 4 = Good demonstration of Task; 5 = Outstanding demonstration of Task.

Task	0	1	2	3	4	5
Wear proper safety equipment.						
Cut the material being used for the panel into pieces that are about an inch longer than the finished panel will be.						
Alternating the growth rings, arrange the boards into a pleasing pattern, and mark it for easy reassembly later.						
Joint the edges of each board, alternating faces against the jointer fence.						
Reassemble the panel and check for flush joints.						
Mark each board to indicate where the tongue or groove will be milled.						
Using a scrap that is the same thickness as the panel material, make a mark in the center of its thickness.						
At the table saw, set the fence so it will cut just to the outside of your mark, and raise the blade to just slightly over 1/4".						
Run your test piece with one face against the fence, and then flip it around and run the other face against the fence.						
Check the resulting groove. It should be 1/4" wide and just over 1/4" deep. Make adjustments if necessary, and then run all the pieces that you have marked to get a groove.						
Install a dado blade set for 1/4" wide in the table saw.						
Clamp a sacrificial fence to the fence, and bring it right over next to the blade.						
Set the blade height at 1/4" above the table.						

Continued

Task	0	1	2	3	4	5
Take another test piece the same thickness as the panel material, and run it face down against the fence. Flip it over and run the opposite face; this process creates a tongue						
Check the fit of the test tongue you just made in one of the grooves you made previously. It should fit snugly with a little bit of room in the bottom of the groove.						
Make any adjustments needed, and then run all the pieces that are marked to get a tongue.						
Glue and clamp up the panel.						
TOTAL POINTS						
FINAL (AVERAGED) SCORE						

EVALUATOR NOTES	STUDENT NOTES

280

PROCEDURE 20-1: Building a Face Frame Using Half-Lap Joints

Objective: The student shall successfully demonstrate building a face frame using half-lap joints.

Materials Needed: Material for stiles, rails, and mullions if included in plan
Table saw
Dado blade
Miter gauge
Measuring tool
Square
Pencil
Chisel
Glue
Clamps

Safety Requirements: Eye protection
Hearing protection

References: *Chapter 20, Cabinets and Casework*
Procedure 20-1, p. 511

Student Name: _____ **Date:** _____

Evaluator Name: _____ **Date:** _____

Ratings: 0 = Skipped Task; 1 = Attempted, but did not complete Task; 2 = Poor demonstration of Task;
3 = Fair demonstration of Task; 4 = Good demonstration of Task; 5 = Outstanding demonstration of Task.

Task	0	1	2	3	4	5
Wear proper safety equipment.						
Determine the length of the stiles, rails, and any mullions.						
Cut the stiles and rails to the width specified in the plans and the lengths determined.						
Lay out for the half-lap joints, paying careful attention to get the marks on the correct side of each board.						
Install a dado blade set for its widest setting on the table saw, and raise it so that its height is equal to half the thickness of the wood you are using.						
Adjust the fence so that your layout mark is on the far side of the blade.						
Using two test pieces that are the same thickness as the face frame material, support one piece with the miter gauge and make a cut.						
Slide the piece to your left and make successive cuts until you have made a half-lap. Repeat for the second test piece.						
Trim any ridges left from the saw with a chisel.						
Put the two test pieces together to form a corner and check the fit.						

Continued

281

Task	0	1	2	3	4	5
Adjust the height of the blade and the fence as needed. If the fit is not flush, the blade height needs to be adjusted. If the corner is not flush, the fence needs to be adjusted. Continue making test cuts until you are satisfied.						
Machine half-laps on the pieces of the face frame.						
Chisel the face of each joint flush if needed.						
Apply glue to the joints and clamp up the face frame, positioning clamps at each joint and across the frame.						
Check the frame for square, and adjust as needed.						
TOTAL POINTS						
FINAL (AVERAGED) SCORE						

EVALUATOR NOTES	STUDENT NOTES

PROCEDURE 21-1: Making a Raised Panel on the Router Table

Objective: The student shall successfully demonstrate making a raised panel on the router table.

Materials Needed: Solid wood panel
Grooved stiles and rails for the frame desired
Tools to size the panel
Measuring tool
Router table
Router
Raised panel bit

Safety Requirements: Eye protection
Hearing protection

References: *Chapter 21, Cabinet Doors and Drawers*
Procedure 21-1, p. 542

Student Name: _____ **Date:** _____

Evaluator Name: _____ **Date:** _____

Ratings: 0 = Skipped Task; 1 = Attempted, but did not complete Task; 2 = Poor demonstration of Task; 3 = Fair demonstration of Task; 4 = Good demonstration of Task; 5 = Outstanding demonstration of Task.

Task	0	1	2	3	4	5
Wear proper safety equipment.						
Determine the size of the panel by first measuring the inside distance between the two stiles and the two rails.						
Measure the depth of the groove in the stiles and rails, double that measurement, and add it to the length and width determined.						
Subtract 1/16" from the overall length and 1/4" from the width.						
Cut your panel to the dimensions determined.						
Install a raised-panel cutter in the router table, raising it just above the table.						
Align the fence with the router bit bearing.						
Keeping the panel firmly against the fence, rout the end of the panel first.						
Now rout a long side, the other end, and the other long side.						
Raise the bit 1/8" and repeat, again always following an end-grain pass with a long-grain pass.						
Continue until you have the profile desired.						
TOTAL POINTS						
FINAL (AVERAGED) SCORE						

EVALUATOR NOTES	STUDENT NOTES

PROCEDURE 21-2: Milling a Groove-and-Stub Tenon Joint on the Router Table

Objective:	The student shall successfully demonstrate milling a groove-and-stub tenon joint on the router table.
Materials Needed:	Two 3/4" thick boards between 2" and 4" wide
	Measuring tool
	Router table
	Router
	1/4" straight bit
	1/2" or larger straight bit
	Miter gauge
Safety Requirements:	Eye protection
	Hearing protection
References:	*Chapter 21, Cabinet Doors and Drawers*
	Procedure 21-2, p. 546

Student Name: _____ Date: _____

Evaluator Name: _____ Date: _____

Ratings: 0 = Skipped Task; 1 = Attempted, but did not complete Task; 2 = Poor demonstration of Task; 3 = Fair demonstration of Task; 4 = Good demonstration of Task; 5 = Outstanding demonstration of Task.

Task	0	1	2	3	4	5
Wear proper safety equipment.						
Insert a 1/4" straight bit in the router table and set the fence so that the groove will be milled in the center of the stock.						
Raise the bit 1/4" above the table, and rout a groove on the edge of a test piece.						
Check to see if it is centered on the piece, and adjust the fence if necessary.						
Run grooves on one edge of both pieces. If they need to be deeper than 1/4", raise the bit gradually between passes until the desired depth of cut is achieved.						
Replace the router bit with a 1/2" or larger straight bit.						
Using one of the grooved pieces as a guide, set the depth of cut so that it will take away material up to the groove.						
Adjust the fence so that the amount of bit exposed is equal to the depth of the groove.						
Check to ensure that the router table fence is parallel to the miter groove by measuring from the fence to the miter gauge groove at several points.						
Supporting a test piece with a miter gauge, run one end over the bit.						

Continued

Task	0	1	2	3	4	5
Flip the piece over and run the same end again.						
Test the fit of the test piece. If it is too thick, raise the bit slightly, remembering that the amount you raise it is doubled when you make the cut. If the tongue is too thin, lower the bit. If it is too long, move the fence in to cover more of the bit. If it is too short, move the fence back. Continue testing until you have a good fit.						
Mill the stub tenon in the finished piece.						
TOTAL POINTS						
FINAL (AVERAGED) SCORE						

EVALUATOR NOTES	STUDENT NOTES

286

PROCEDURE 21-3: Milling a Rabbet-and-Dado Joint on the Router Table

Objective: The student shall successfully demonstrate milling a rabbet-and-dado joint on the router table.

Materials Needed: Two boards of equal thickness
Scrap of the same thickness as the boards for tests
Measuring tool
Router table
Router
Straight bits
Miter gauge

Safety Requirements: Eye protection
Hearing protection

References: *Chapter 21, Cabinet Doors and Drawers*
Procedure 21-3, p. 553

Student Name: _____ Date: _____

Evaluator Name: _____ Date: _____

Ratings: 0 = Skipped Task; 1 = Attempted, but did not complete Task; 2 = Poor demonstration of Task; 3 = Fair demonstration of Task; 4 = Good demonstration of Task; 5 = Outstanding demonstration of Task.

Task	0	1	2	3	4	5
Wear proper safety equipment.						
Install a straight bit in the router table; the bit diameter should be greater than half the thickness of the stock you are working with.						
Raise the bit so that its height is half the thickness of the stock you are using.						
Set the fence so that the amount of bit exposed is half the thickness of the stock.						
Make sure that the miter gauge groove and the fence are parallel to each other by measuring the distance between them at several points.						
Using the miter gauge to support a test piece, mill a rabbet.						
Check the result to see if its length and thickness are half the total stock thickness, and adjust the bit height and fence as necessary.						
Mill a rabbet on the finish board.						
Replace the bit with another straight bit whose diameter is equal to half the thickness of the stock being used.						
Raise the bit so that its height is half the thickness of the stock.						
Set the fence so the distance between the fence and the bit is half the thickness of the stock.						

Continued

Task	0	1	2	3	4	5
Supporting a test piece with the miter gauge, machine a dado.						
Check the fit of the test piece with the final rabbeted board.						
Adjust the bit height and fence position until you get a good flush fit.						
Machine a dado on the finish board.						
TOTAL POINTS						
FINAL (AVERAGED) SCORE						

EVALUATOR NOTES	STUDENT NOTES

PROCEDURE 21-4: Machining a Sliding Dovetail

Objective: The student shall successfully demonstrate machining a sliding dovetail.

Materials Needed: Two boards
Scrap of the same thickness as the boards for tests
Measuring tool
Router table
Router
Straight bit
Dovetail bit
Miter gauge

Safety Requirements: Eye protection
Hearing protection

References: *Chapter 21, Cabinet Doors and Drawers*
Procedure 21-4, p. 557

Student Name: _____ **Date:** _____

Evaluator Name: _____ **Date:** _____

Ratings: 0 = Skipped Task; 1 = Attempted, but did not complete Task; 2 = Poor demonstration of Task; 3 = Fair demonstration of Task; 4 = Good demonstration of Task; 5 = Outstanding demonstration of Task.

Task	0	1	2	3	4	5
Wear proper safety equipment.						
Mill a dado in one board using a straight bit installed in the router table. The depth of the dado should be half the stock thickness, and its width should be 1/8" to 1/4" less than the thickness of the piece that will get the dovetail.						
Insert a dovetail bit into a table-mounted router, and set its height to be equal to the depth of the dado just machined.						
Set the router table fence so that the cut will be made in the center of the dado.						
Make sure that the fence is parallel to the miter gauge slot.						
Supporting the dadoed piece with the miter gauge, run it over the dovetail bit, creating a dovetailed dado.						
Remove the miter gauge from the table and loosen the fence, but keep the bit at the same height.						
Adjust the fence so that most of the bit is covered, and then run a test piece in the upright position.						
Flip it around and run the other side. You now have a dovetail.						
Check the test piece in the dado dovetail. If it is too thick and won't go in, move the fence to expose more of the bit. If it is too thin and loose, move the fence in, covering more of the bit.						
Run successive test cuts until you have a good fit.						
Mill the dovetail in the finish board.						
TOTAL POINTS						
FINAL (AVERAGED) SCORE						

EVALUATOR NOTES	STUDENT NOTES

PROCEDURE 22-1: Making and Using a Fixed Tapering Jig for a Two-Sided Taper

Objective: The student shall successfully demonstrate making and using a fixed tapering jig for a two-sided taper.

Materials Needed: 3/4" plywood for jig parts
Measuring tool
Pencil
Drill, driver, and screws for fastening jig together
Two toggle clamps
Square stock for tapering
Table saw
Rip or combination blade

Safety Requirements: Eye protection
Hearing protection

References: *Chapter 22, Tables and Desks*
Procedure 22-1, p. 570

Student Name: _____ Date: _____

Evaluator Name: _____ Date: _____

Ratings: 0 = Skipped Task; 1 = Attempted, but did not complete Task; 2 = Poor demonstration of Task; 3 = Fair demonstration of Task; 4 = Good demonstration of Task; 5 = Outstanding demonstration of Task.

Task	0	1	2	3	4	5
Wear proper safety equipment.						
Review the diagram given in your textbook and cut the base, guide bar, and stop to size.						
Lay out the desired taper on one of the pieces you will be tapering.						
Place the piece just laid out on the base so that the part to be tapered overhangs the side; draw a line along the side that rests on the base and across the end.						
Fasten the guide bar to the base along the long line you just drew, and fasten the stop to the short line.						
Screw toggle clamps to the guide bar, taking care that when they are in the down position, they will not be in the cut line.						
Set the fence at 8 inches.						
Clamp the piece to be tapered in the jig, and raise the blade high enough to clear both the jig and the piece being tapered.						
Turn on the saw and guide the jig along the fence to make the first tapered cut.						
Unclamp the work piece and turn it 90°. Re-clamp it in the jig, using the cutoff from the first cut as a shim to keep the clamps tight.						
Turn on the saw and cut the second taper.						
TOTAL POINTS						
FINAL (AVERAGED) SCORE						

EVALUATOR NOTES	STUDENT NOTES

PROCEDURE 22-2: Making Button Blocks

Objective:	The student shall successfully demonstrate making button blocks.
Materials Needed:	Table needing button blocks or plan for same
	3/4" × 4"(minimum) × 12" hardwood board
	Jointer
	Miter saw
	Table saw
	Dado blade
	Measuring tool
	Pencil
	Square
	Band saw and fence
	Drill press or portable drill
	Drill bit
	Countersink bit
Safety Requirements:	Eye protection
	Hearing protection
	Push stick
References:	*Chapter 22, Tables and Desks*
	Procedure 22-2, p. 578

Student Name: _____ Date: _____

Evaluator Name: _____ Date: _____

Ratings: 0 = Skipped Task; 1 = Attempted, but did not complete Task; 2 = Poor demonstration of Task; 3 = Fair demonstration of Task; 4 = Good demonstration of Task; 5 = Outstanding demonstration of Task.

Task	0	1	2	3	4	5
Wear proper safety equipment.						
Joint one edge and square cut both ends of the board.						
Using the table saw and a dado blade, mill a rabbet that is 1/2" wide on both ends of your stock. The depth of the cut should be such that the portion of material remaining will slide into the groove milled in the apron.						
Measure over 2" from each end of the board and make layout lines for crosscutting later.						
Set the band saw fence for 1", and rip off individual blocks. Use a push stick.						
Drill and countersink a hole through the center of the solid part of each block. The hole should be large enough that the screw being used to attach the top will drop through it easily.						
Cut blocks to length along the crosscut lines you marked out earlier.						
TOTAL POINTS						
FINAL (AVERAGED) SCORE						

EVALUATOR NOTES	STUDENT NOTES

PROCEDURE 22-3: Making an Eight-Sided Column

Objective:	The student shall successfully demonstrate making an eight-sided column.
Materials Needed:	Eight boards for the column
	Paper or plywood larger than the desired perimeter of the column
	Pencil
	Compass, trammel points, or beam compass
	Straightedge
	Measuring tool
	Miter saw
	Table saw
	Glue
	Wide painter's tape
	Band clamps
Safety Requirements:	Eye protection
	Hearing protection
	Push stick
References:	*Chapter 22, Tables and Desks*
	Procedure 22-3, p. 583

Student Name: _____ Date: _____

Evaluator Name: _____ Date: _____

Ratings: 0 = Skipped Task; 1 = Attempted, but did not complete Task; 2 = Poor demonstration of Task; 3 = Fair demonstration of Task; 4 = Good demonstration of Task; 5 = Outstanding demonstration of Task.

Task	0	1	2	3	4	5
Wear proper safety equipment.						
On a large sheet of paper or a scrap piece of plywood, draw a straight line, set a compass or trammel points in the center of the line, and draw a circle that is the diameter of the column you want to make.						
Label the center of the circle 1.						
Label the two points where the circle intersects the lines 2 and 3.						
Open the compass a bit wider than the radius used to draw the circle; place its point at point 2 and then at point 3 and scribe short arcs that intersect above and below the circle. Label these intersections points A and B.						
Connect points A and B with a straight line.						
Label the points where this line touches the circle points 4 and 5.						
Set the compass opening for the radius of the circle again and place its point at point 2 and scribe short arcs outside the circle; do the same at points 3, 4, and 5. The intersections of these lines create points to be labeled C, D, E, and F.						

Continued

Task	0	1	2	3	4	5
Connect points C and E and points D and F.						
Draw lines between the points along the perimeter of the circle to establish the octagon.						
Measure one of these lines to learn the width of each of the eight pieces that will make up the column.						
Crosscut eight pieces to the desired length.						
Rip the pieces slightly wider than the measurement obtained earlier.						
Tilt the table saw blade to 22.5°.						
Using the fence as a guide, rip this bevel on one side of each of your eight pieces.						
Set the fence so that the distance between the fence and the blade is the determined width. It will be easiest to do this if you mark one of your pieces, measuring from the bevel just cut.						
Rip the opposing bevel on each of your pieces.						
Glue and secure the column with wide painter's tape and band clamps.						
Remove the band clamps and clean up any glue residue.						
TOTAL POINTS						
FINAL (AVERAGED) SCORE						

EVALUATOR NOTES	STUDENT NOTES

PROCEDURE 23-1: Applied Base

Objective:	The student shall successfully demonstrate making and installing an applied base to a chest.
Materials Needed:	Chest
	Material for base
	Pencil
	Measuring tool
	Miter saw
	Router
	Desired profile bit
	Glue
	Finish nails
Safety Requirements:	Eye protection
	Hearing protection
References:	*Chapter 23, Chests*
	Procedure 23-1, p. 607

Student Name: _____ Date: _____

Evaluator Name: _____ Date: _____

Ratings: 0 = Skipped Task; 1 = Attempted, but did not complete Task; 2 = Poor demonstration of Task; 3 = Fair demonstration of Task; 4 = Good demonstration of Task; 5 = Outstanding demonstration of Task.

Task	0	1	2	3	4	5
Wear proper safety equipment.						
Rout a profile on one edge of a board that will serve as a baseboard to a chest. The board or boards being used must be long enough to go around the perimeter of the chest. Allow some extra material.						
Miter the board just profiled to go around the bottom of the chest.						
If the bottom is set into a groove in the sides, and the front is open, glue and nail a cleat under the bottom for attaching the baseboard.						
Glue and nail the mitered baseboard in place.						
TOTAL POINTS						
FINAL (AVERAGED) SCORE						

EVALUATOR NOTES	STUDENT NOTES

PROCEDURE 23-2: Making a Scrolled Base

Objective: The student shall successfully demonstrate making and installing a scrolled base to a chest.

Materials Needed: Chest
Material for base
Pencil
Measuring tool
Miter saw
Band saw or jigsaw
Table saw and dado blade or router and rabbeting bit
Glue
Finish nails

Safety Requirements: Eye protection
Hearing protection

References: *Chapter 23, Chests*
Procedure 23-2, p. 609

Student Name: _____ **Date:** _____

Evaluator Name: _____ **Date:** _____

Ratings: 0 = Skipped Task; 1 = Attempted, but did not complete Task; 2 = Poor demonstration of Task; 3 = Fair demonstration of Task; 4 = Good demonstration of Task; 5 = Outstanding demonstration of Task.

Task	0	1	2	3	4	5
Wear proper safety equipment.						
Select enough 3/4"-thick material to go around the perimeter of the case, allowing some extra.						
Lay out a pattern on the base, and cut it on the band saw or with a jigsaw.						
Machine a 3/8" × 3/8" rabbet in the inside edge of each of the base pieces.						
Turn the case upside down on your bench, and miter the base pieces to fit around the case, so that the rabbet supports the case.						
Glue and nail the mitered pieces together.						
Cut glue blocks to go in each corner of the base. They may be triangular or square.						
Glue and fasten the glue blocks in place.						
Spread glue in the rabbet on the base and set the chest in the base.						
TOTAL POINTS						
FINAL (AVERAGED) SCORE						

EVALUATOR NOTES	STUDENT NOTES

PROCEDURE 23-3: Making an Ogee Bracket Foot

Objective:	The student shall successfully demonstrate making an ogee bracket foot.
Materials Needed:	Template material—paper and/or 1/4" ply or hardboard
	Piece of posterboard for cove template
	Material for foot; 1 1/2" × 1 1/2" × 16" stock
	1/8" hardboard for spline
	Straight stock for angled table saw fences
	Pencil
	Measuring and layout tools
	Band saw
	Table saw
	Crosscut or combination blade
	Dado blade
	Glue
	Hand plane
	Sandpaper
Safety Requirements:	Eye protection
	Hearing protection
References:	*Chapter 23, Chests*
	Procedure 23-3, p. 613

Student Name: _____   Date: _____

Evaluator Name: _____   Date: _____

Ratings: 0 = Skipped Task; 1 = Attempted, but did not complete Task; 2 = Poor demonstration of Task;
3 = Fair demonstration of Task; 4 = Good demonstration of Task; 5 = Outstanding demonstration of Task.

Task	0	1	2	3	4	5
Wear proper safety equipment.						
Using the diagram in your textbook as a guide, make a template for the foot you want to make.						
Copy the shape of the ogee on both ends of a 1 1/2" × 1 1/2" × 16" board, using the short part of the template as a guide.						
Raise the table saw blade to the full height of the cove. Mark the points where the blade enters and exits the table.						
Make a posterboard template with a window whose inside width is the same as the cove's final width. Place the template on the saw table and angle it so that the long edges touch the marks you made in the previous step.						
Mark one of the long inside edges of the template and remove it						
Clamp a straight piece of stock to the mark just made on the table saw to serve as one fence.						
Place the material to be machined against the fence and set a second fence.						

Continued

301

Task	0	1	2	3	4	5
Lower the blade so it is only about 1/16" above the table, and then cut the cove by making a series of light passes, raising the blade between passes until the desired depth of cove is reached.						
Elongate the cove you cut using a dado blade set to a width, height, and angle that will cut up near your layout line.						
Rip the blank to final width, using your layout mark as a guide.						
Rough out the roundover located on the top outside corner with a regular blade on the table saw, tilted to 45°.						
Plane and sand your foot blank to its final shape, using the ogee layout on each end as a guide.						
Miter the blank into the two halves of the foot.						
Cut a groove for a spline.						
Use the long side of the template to draw the bracket outline on the flat side of each piece.						
Cut out the shape on the band saw.						
Glue and clamp the two parts of the foot together with a spline in place. A glue block attached to the back of the miter will help keep the assembly square and add strength.						
Remove the clamps, clean up any glue residue, and do your final sanding.						
TOTAL POINTS						
FINAL (AVERAGED) SCORE						

EVALUATOR NOTES	STUDENT NOTES

PROCEDURE 24-1: Installing Bed Bolts

Objective: The student shall successfully demonstrate installing bed bolts.

Materials Needed: Bed post or material of similar width and thickness
                        Bed rail or material of similar width and thickness
                        Bed bolt and nut
                        Square
                        Pencil
                        Mortising machine, drill press and chisels or router and straight bit to mill
                        mortises
                        Table saw
                        Dado blade
                        Crosscut or combination blade
                        Drill press
                        1" flat-bottomed bit
                        3/8" bit
                        Portable drill
                        Long 3/8" bit

Safety Requirements: Eye protection
                        Hearing protection

References: *Chapter 24, Beds*
                        Procedure 24-1, p. 628

Student Name: _____   Date: _____

Evaluator Name: _____   Date: _____

Ratings: 0 = Skipped Task; 1 = Attempted, but did not complete Task; 2 = Poor demonstration of Task;
3 = Fair demonstration of Task; 4 = Good demonstration of Task; 5 = Outstanding demonstration of Task.

Task	0	1	2	3	4	5
Wear proper safety equipment.						
Mill a centered mortise that is 1" wide by 5" long and 9/16" deep in the post.						
Make 1/2" long tenons on the rails to match the mortises.						
Mark the horizontal centerline of the mortise in each post and transfer it around to the opposite side of the post. Then mark the intersecting vertical centerline.						
Take the post to the drill press and drill a 1" counterbore 3/4" deep at the marked location.						
Drill 3/8" holes centered in the counterbore the rest of the way through the post.						
Clamp the rails to the post and, using a long 3/8" bit, drill three 3/4" into the rail.						
Use a square and pencil to carry the centerline of the 3/8" hole just drilled around to the inside of the rail; measure over and mark two 3/4" from the shoulder of the tenon.						

Continued

Task	0	1	2	3	4	5
Use a 1" Forstner bit to drill a hole 1 1/4" deep at the centerline just marked.						
Insert the bed bolt through the post and into the tenon.						
Place the bed bolt nut in the 1" hole drilled in the rail, and tighten it, securely attaching the post to the rail.						
Cover the holes in the posts with purchased bed bolt covers, or make your own.						
TOTAL POINTS						
FINAL (AVERAGED) SCORE						

EVALUATOR NOTES	STUDENT NOTES

304

PROCEDURE 24-2: Installing Mortised Bed Rail Fasteners

Objective: The student shall successfully demonstrate installing mortised bed rail fasteners.

Materials Needed: Bed post or material of similar width and thickness
Bed rail or material of similar width and thickness
Bed rail fastener set and screws
Knife
Pencil
Chisels
Mallet
Drill
Self-centering bit
Screwdriver
1/4" drill bit

Safety Requirements: Eye protection
Hearing protection

References: *Chapter 24, Beds*
Procedure 24-2, p. 635

Student Name: _____ Date: _____

Evaluator Name: _____ Date: _____

Ratings: 0 = Skipped Task; 1 = Attempted, but did not complete Task; 2 = Poor demonstration of Task;
3 = Fair demonstration of Task; 4 = Good demonstration of Task; 5 = Outstanding demonstration of Task.

Task	0	1	2	3	4	5
Wear proper safety equipment.						
Center the male bracket both horizontally and vertically on the end of the rail to which it will be fastened and mark around it.						
With a chisel and mallet, mortise out the marked area so that the plate of the bracket will sit flush with the end of the rail.						
Place the bracket in position on the end of the rail and drill pilot holes.						
Screw the bracket in place.						
Position the female bracket in the desired location on the post and mark around it.						
Chisel out a recess for the plate of this bracket in the post.						
Place the female bracket in the recess in the post, and mark out the location for the hook mortises.						
Remove the bracket and use a 1/4" bit to drill overlapping holes 1/2" deep in the two marked locations.						
Finish the mortises with a chisel.						
Set the female bracket in position on the post and drill pilot holes.						
Screw the bracket in place.						
TOTAL POINTS						
FINAL (AVERAGED) SCORE						

EVALUATOR NOTES	STUDENT NOTES

PROCEDURE 26-1: Installing a Butt Hinge

Objective:	The student shall successfully demonstrate installing a butt hinge.
Materials Needed:	Butt hinge and screws
	2 boards to fasten hinge to
	Knife
	Chisels
	Mallet
	Drill
	Self-centering bit
	Screwdriver
Safety Requirements:	Eye protection
	Hearing protection
References:	*Chapter 26, Hardware*
	Procedure 26-1, p. 679

Student Name: _____ Date: _____

Evaluator Name: _____ Date: _____

Ratings: 0 = Skipped Task; 1 = Attempted, but did not complete Task; 2 = Poor demonstration of Task; 3 = Fair demonstration of Task; 4 = Good demonstration of Task; 5 = Outstanding demonstration of Task.

Task	0	1	2	3	4	5
Wear proper safety equipment.						
Lay one leaf of the hinge on the piece that will receive it, and trace around it with a sharp knife.						
Use a knife to mark the depth of the hinge on the edge of the board.						
With a chisel and mallet, deepen the knife marks made in the first step. The bevel of the chisel should face inward toward the material that will be removed.						
Use a chisel and mallet to make a series of cuts 1/8" to 1/4" apart inside the area just outlined.						
Use a chisel to remove the material to the depth marked.						
Check the fit of the leaf and adjust as needed.						
Use a self-centering bit to drill pilot holes at the screw locations, using the hinge as a guide.						
Repeat the previous steps for the other leaf of the hinge.						
Attach each leaf of the hinge to the mating recesses with the screws provided.						
TOTAL POINTS						
FINAL (AVERAGED) SCORE						

EVALUATOR NOTES	STUDENT NOTES

PROCEDURE 26-2: Installing Formed Hinges

Objective:	The student shall successfully demonstrate installing formed hinges.
Materials Needed:	Pair of formed hinges and screws
	Door and cabinet or two boards 3/4" thick
	Knife
	Chisels
	Mallet
	Drill
	Self-centering bit
	Screwdriver
Safety Requirements:	Eye protection
	Hearing protection
References:	*Chapter 26, Hardware*
	Procedure 26-2, p. 681

Student Name: _____   Date: _____

Evaluator Name: _____   Date: _____

Ratings: 0 = Skipped Task; 1 = Attempted, but did not complete Task; 2 = Poor demonstration of Task;
3 = Fair demonstration of Task; 4 = Good demonstration of Task; 5 = Outstanding demonstration of Task.

Task	0	1	2	3	4	5
Wear proper safety equipment.						
Place the part of the hinge that attaches to the back of the door in the place indicated on the plans.						
Holding the hinge in place, use a self-centering bit to drill the screw holes for the screws that will attach the plate to the back of the door.						
Attach the hinges to the back of the door using the screws provided.						
With the hinges attached to the door, open the hinges and place the door in the opening of the cabinet.						
Adjust the door up and down so that you have an equal space at the bottom and top of the door.						
Hold the door in position; use a self-centering drill bit to drill pilot holes for the top screw for each hinge, placing them in the center of the elongated hole.						
Screw the hinges to the cabinet through the pilot holes just drilled, and close the door to check the top and bottom clearances.						
If the hinges need to be adjusted, loosen the screw just inserted, shift the door to the right position, and then retighten the screws.						
When you are satisfied with the position of the door relative to the cabinet, drill pilot holes for the lower screw in each hinge, positioning them at the top of the elongated hole.						
Insert the remaining two screws.						
TOTAL POINTS						
FINAL (AVERAGED) SCORE						

EVALUATOR NOTES	STUDENT NOTES

PROCEDURE 26-3: Making Multiple Simple Wooden Pulls

Objective: The student shall successfully demonstrate making multiple simple wooden pulls.

Materials Needed: Material for pulls
Jointer
Table saw
Combination blade
Router
Router table
3/4" cove bit
1/4" roundover bit
Miter saw

Safety Requirements: Eye protection
Hearing protection

References: *Chapter 26, Hardware*
Procedure 26-3, p. 691

Student Name: _____ Date: _____

Evaluator Name: _____ Date: _____

Ratings: 0 = Skipped Task; 1 = Attempted, but did not complete Task; 2 = Poor demonstration of Task; 3 = Fair demonstration of Task; 4 = Good demonstration of Task; 5 = Outstanding demonstration of Task.

Task	0	1	2	3	4	5
Wear proper safety equipment.						
Mill the material you wish to make into pulls 3/4" thick, at least 3" wide, and 22" long. Both edges should be jointed.						
Install a 3/4" cove bit in the router table and in a series of passes, machine a 1/2"-deep cove on both edges of the material						
Rip two strips from the piece you just routed to 1 1/4" and then make a light jointer pass to clean up any saw marks.						
Install a 1/4" roundover bit in the router table and round over the edges.						
Cut the strips into 4" lengths. You will have 10 pulls since you started with a 22" long blank.						
TOTAL POINTS						
FINAL (AVERAGED) SCORE						

EVALUATOR NOTES	STUDENT NOTES

PROCEDURE 26-4: Installing Recessed Pilasters

Objective: The student shall successfully demonstrate installing recessed pilasters.
Materials Needed: Pilaster(s)
                          Ring shank nails or screws
                          3/4" sheet stock to install pilaster
                          Table saw and dado blade or router and 5/8" straight bit
                          Measuring tool
                          Hammer or screwdriver
Safety Requirements: Eye protection
                          Hearing protection
References: *Chapter 26, Hardware*
                          Procedure 26-4, p. 696

Student Name: _____   Date: _____

Evaluator Name: _____   Date: _____

Ratings: 0 = Skipped Task; 1 = Attempted, but did not complete Task; 2 = Poor demonstration of Task;
3 = Fair demonstration of Task; 4 = Good demonstration of Task; 5 = Outstanding demonstration of Task.

Task	0	1	2	3	4	5
Wear proper safety equipment.						
Mill dadoes 5/8" wide and 3/16" deep at the locations where the pilaster(s) will be inserted.						
Press the pilasters into place.						
Secure the pilaster(s) with ring-shank nails or screws through the holes milled in the pilaster(s)						
TOTAL POINTS						
FINAL (AVERAGED) SCORE						

EVALUATOR NOTES	STUDENT NOTES

PROCEDURE 29-1: Determining Kerf Spacing for a Given Bend

Objective: The student shall successfully demonstrate determining kerf spacing for a given bend.

Materials Needed: Board the same thickness as the material to be bent
Miter saw with depth stop or table saw
Measuring tool
Pencil
Clamp

Safety Requirements: Eye Protection
Hearing Protection

References: *Chapter 29, Bending Wood*
Procedure 29-1, p. 751

Student Name: _____ Date: _____

Evaluator Name: _____ Date: _____

Ratings: 0 = Skipped Task; 1 = Attempted, but did not complete Task; 2 = Poor demonstration of Task; 3 = Fair demonstration of Task; 4 = Good demonstration of Task; 5 = Outstanding demonstration of Task.

Task	0	1	2	3	4	5
Wear proper safety equipment.						
Cut a piece of wood as thick as the piece you want to bend.						
Cut a kerf about 6" from the end of the board and deep enough so that 1/8" of material is left.						
Measure and mark the length of the radius of the desired curve from the kerf.						
Clamp the end of the board to your bench, and lift up the other end until the kerf closes.						
Measure the distance between the board and the bench at the radius mark. This distance is equal to the minimum kerf spacing.						
TOTAL POINTS						
FINAL (AVERAGED) SCORE						

EVALUATOR NOTES	STUDENT NOTES

PROCEDURE 29-2: Build a Steam Box

Objective: The student shall successfully build a steam box.

Materials Needed: 3/4" exterior plywood for the box
2 × 4 material for legs
3/8" dowel stock
Meat thermometer (optional)
Butt hinge
Screen door hook and eye
Galvanized screws
Water resistant or waterproof glue
Table saw or circular saw
Drill and driver
3/8" bit
1/4" bit
Jigsaw
Measuring and layout tools
Pencil

Safety Requirements: Eye Protection
Hearing Protection

References: *Chapter 29, Bending Wood*
Procedure 29-2, p. 757

Student Name: _____ Date: _____

Evaluator Name: _____ Date: _____

Ratings: 0 = Skipped Task; 1 = Attempted, but did not complete Task; 2 = Poor demonstration of Task; 3 = Fair demonstration of Task; 4 = Good demonstration of Task; 5 = Outstanding demonstration of Task.

Task	0	1	2	3	4	5
Wear proper safety equipment.						
Refer to your textbook for a diagram of the box. This box is 56" long, but you can adjust the length to suit your needs.						
Cut the pieces needed to size.						
Line up the sides of the box and drill 3/8" holes through both pieces.						
Assemble the sides, the bottom, and one end of the box with galvanized screws.						
Tap the 3/8" dowels through the sides of the box.						
Glue and screw the top on the box.						
On the bottom of the box in the center, drill or cut a hole to accommodate the pipe that will bring the steam into the box.						
Glue the plug to the door. Check your box opening; the plug should be 1/8" shorter in length and width than the opening to allow for swelling.						

Continued

317

Task	0	1	2	3	4	5
Attach the door to the box with a large butt hinge.						
Attach a screen-door hook and eye to keep the end closed.						
Turn the box over and drill two 1/4" holes toward one end of the box to allow condensation to drain.						
Drill a tight-fitting hole to accept a meat thermometer on the top of the box near one end if you are going to install a thermometer in your steam box.						
Attach the legs to the box with galvanized screws in accordance with the diagram in your text. This tilts the box so that condensation will run out the condensation hole.						
TOTAL POINTS						
FINAL (AVERAGED) SCORE						

EVALUATOR NOTES	STUDENT NOTES

318

PROCEDURE 29-3: Creating a Curve Using Segment Lamination

Objective: The student shall successfully demonstrate creating a curve using segment lamination.

Materials Needed: Heavy paper or posterboard for curve template
Straightedge
Protractor or sliding T-bevel
Measuring tool
Pencil
3/4" stock for segments
Glue
Table saw
Miter saw
Band saw
Scraper
Sander and sandpaper
Veneer (optional)

Safety Requirements: Eye Protection
Hearing Protection

References: *Chapter 29, Bending Wood*
Procedure 29-3, p. 767

Student Name: _____ **Date:** _____

Evaluator Name: _____ **Date:** _____

Ratings: 0 = Skipped Task; 1 = Attempted, but did not complete Task; 2 = Poor demonstration of Task; 3 = Fair demonstration of Task; 4 = Good demonstration of Task; 5 = Outstanding demonstration of Task.

Task	0	1	2	3	4	5
Wear proper safety equipment.						
Decide on the curve you wish to create, and make a full-size template of the inside and outside of the curve.						
Draw a straight line along the base of the curve, and mark its center point.						
Rip material to width, making it 1" wider than the thickness of your curve.						
Determine the rough length of individual pieces by setting a piece on your template and noting how much is required to cover both the inside and outside of the curve.						
Determine the angle that will be cut on the ends of each piece by first marking off the rough length along the template curve.						
Draw lines from the length marks you made along the curve back to the center point marked on the horizontal line.						

Continued

Task	0	1	2	3	4	5
Use a protractor or sliding T-bevel placed on the horizontal line to determine the angle.						
Divide the angle in half to determine the saw setting for the cut.						
Cut the individual pieces.						
Assemble the first row of pieces.						
Spread glue on the first row and lay the second row in place, staggering them by half.						
Repeat for as many rows as are needed to achieve the desired height.						
Place weight on top of the assembly and wait for it to dry.						
Transfer the curve from the template onto the glued-up assembly.						
Cut the curve on the band saw.						
Scrape and sand smooth.						
Veneer the outside of the curve if desired.						
TOTAL POINTS						
FINAL (AVERAGED) SCORE						

EVALUATOR NOTES	STUDENT NOTES

PROCEDURE 31-1: Making and Using a Scratch Stock

Objective:	The student shall successfully demonstrate making and using a scratch stock.
Materials Needed:	3/4" thick hardwood, approximately 2 1/2" × 7"
	Bolt and nut
	An old band saw blade, handsaw, hacksaw, or scraper
	Measuring tool
	Layout tools
	Band saw
	Files
	Drill press
	Bit of the same size as the bolt
Safety Requirements:	Eye Protection
	Hearing Protection
References:	*Chapter 31, Decorative Techniques*
	Procedure 31-1, p. 803

Student Name: _____   Date: _____

Evaluator Name: _____   Date: _____

Ratings: 0 = Skipped Task; 1 = Attempted, but did not complete Task; 2 = Poor demonstration of Task;
3 = Fair demonstration of Task; 4 = Good demonstration of Task; 5 = Outstanding demonstration of Task.

Task	0	1	2	3	4	5
Wear proper safety equipment.						
Using the band saw, cut a stopped kerf 2 1/2" long, centered on the thickness of the wood.						
Consult your textbook and cut the block to match the dimensions given.						
Use a file to round the fence of the scratch stock.						
Drill a hole to take the bolt that will secure the cutter.						
Use a piece of an old band saw blade, handsaw, hacksaw, or scraper to make the cutter.						
Lay out the negative of the shape you want to make on the cutter and then shape it with a small file.						
Sharpen the cutter by honing both sides on a stone.						
Secure the cutter in the scratch-stock body and tighten the bolt.						
Draw the scratch stock back and forth in one spot to make a groove or shape.						
TOTAL POINTS						
FINAL (AVERAGED) SCORE						

EVALUATOR NOTES	STUDENT NOTES

PROCEDURE 31-2: Making and Using a Parquetry Jig

Objective: The student shall successfully demonstrate making and using a parquetry jig.

Materials Needed: 3/4" × 24" × 24" MDF, plywood, or particle board
1/2" × 1 1/2" × 24" piece of hardwood
1/4" thick material for spacer blocks
Drill, bit, and driver
Screws
Measuring tool
Pencil
Metal straightedge
Veneer saw or knife

Safety Requirements: Eye Protection
Hearing Protection

References: *Chapter 31, Decorative Techniques*
Procedure 31-2, p. 806

Student Name: _____ Date: _____

Evaluator Name: _____ Date: _____

Ratings: 0 = Skipped Task; 1 = Attempted, but did not complete Task; 2 = Poor demonstration of Task;
3 = Fair demonstration of Task; 4 = Good demonstration of Task; 5 = Outstanding demonstration of Task.

Task	0	1	2	3	4	5
Wear proper safety equipment.						
Screw the 1/2" × 1 1/2" × 24" piece of hardwood flush with one edge of the 3/4" × 24" × 24" piece.						
Mark out two lines at a 90° angle to the fence. These are called *setting-out lines*.						
Cut two 1/4" thick spacer blocks. Their size is determined by the width of the pieces to be cut. For example, if the strips you will be cutting are 2" wide, you will need two 2" spacer blocks.						
To use the jig, first square one edge of the veneer by lining it up on one of the setting-out lines.						
Put the two spacer blocks against the fence and butt the straightedge to them.						
Keeping the veneer on the setting-out line, slide it under the straightedge so that it shows, but not all the way to the fence.						
Cut it with a veneer saw or a knife by running it along the straightedge.						
To make a series of strips of the same width, slide the squared-up edge all the way to the fence, keeping it aligned with the setting-out line.						
Butt the spacer blocks against the fence and slide the straightedge up against them.						
Cut and repeat as many times as needed.						
TOTAL POINTS						
FINAL (AVERAGED) SCORE						

323

EVALUATOR NOTES	STUDENT NOTES